T0336815

Power Electronics Applied to Industrial Systems and Transports 4

*Series Editor*
*Bernard Multon*

# Power Electronics Applied to Industrial Systems and Transports

*Volume 4*
*Electromagnetic Compatibility*

Nicolas Patin

First published 2015 in Great Britain and the United States by ISTE Press Ltd and Elsevier Ltd



ISTE Press Ltd
27-37 St George's Road
London SW19 4EU
UK

www.iste.co.uk

Elsevier Ltd
The Boulevard, Langford Lane
Kidlington, Oxford, OX5 1GB
UK

www.elsevier.com



**British Library Cataloguing in Publication Data**
A CIP record for this book is available from the British Library
**Library of Congress Cataloging in Publication Data**
A catalog record for this book is available from the Library of Congress
ISBN 978-1-78548-003-4

Printed and bound in the UK and US

# Contents

# Preface

Volume 4 in this book series presents a damaging consequence of switching in converters used in power electronics, studied in the context of a domain known as "electromagnetic compatibility" (EMC). In Volume 1 [PAT 15a], we saw that the switching mode used in converters may seem simplistic in terms of degrees of freedom when controlling the power flowing between a power supply and a load. However, this mode was seen to be satisfactory as long as the load presented an inertia sufficient to avoid effects from voltage of current switching. Once this conceptual difficulty is overcome, the gains in terms of energy efficiency and loss reduction (with associated gains regarding the volume and weight of the converter) are considerable. Unfortunately, this is not the full picture as it does not include electromagnetic interference produced by electronic switches in switching mode as switching occurs very quickly and at an increasingly high speed (switching times lower – sometimes much lower – than 1 microsecond and switching frequencies from a few hundred hertz in high power applications to several megahertz in some low power highly miniaturized switch mode power supplies). In these conditions, the great variation in (potentially high) voltages and currents over time results in the production of variable electrical and magnetic fields, which can generate interference in nearby electronic equipment (including subsystems in the converter itself). EMC can be seen as the study of the interference mechanisms which may exist between equipment creating interference (the source) and equipment subject to interference (the victim). Rules for coexistence are established on this basis in order to guarantee successful operation of elements in proximity to one another. This volume will not focus on the standardization approach (which will, nevertheless, be mentioned in Chapter 1), but will concentrate on the study of disturbance mechanisms and tools used to combat these difficulties.

Sources of interference will be presented in Chapter 1, including artificial sources (such as electronic switches in switching mode) but also natural interference (lightning and static electricity carried by the human body). Clearly, the key element in this chapter will be the pulse width modulation (PWM) waveform, which is the most common source of interference in an electronic power converter. Detailed consideration will, therefore, be given to spectral modeling of the PWM waveform using an innovative approach, not widely used in power electronics, based on the Heisenberg uncertainty principle; this principle is widely used in quantum mechanics and signal theory to analyze the duality between notions of temporal and frequency dispersion of a signal.

Chapters 2 and 3 will focus on the paths taken by electromagnetic disturbances between the emitter and the receiver. In Chapter 2, conducted interference will be discussed and, more generally, interference using electrical couplings with lumped elements will be presented. In this case, propagation may be modeled using an equivalent electrical diagram (potentially including parasitic capacitances or mutual inductances, along with common impedances in cases where circuits are galvanically connected). In Chapter 3, we will discuss propagation mechanisms for which the spatiotemporal dimension cannot be reduced (except by the introduction of a cascade of elementary electrical circuits to take account of the non-infinite speed of field and/or voltage propagation, traveling through the length of the propagation channel). This context clearly includes the case of radiated interference, although the division between Chapters 2 and 3 does not fully conform to the classic separation of conducted and radiated interference generally used when studying the EMC.

Finally, this volume includes two appendices, also included in the previous volumes. Appendix 1 provides general formulas for electrical engineering, and was included in Volumes 1, 2 and 3 [PAT 15a, PAT 15b, PAT 15c]. In this volume, the appendix is particularly useful with regard to the Maxwell equations. Appendix 2 is concerned with spectral analysis, as presented in Volume 2. The Fourier transform, in particular, is an important tool used in Chapter 1 of this volume.

Nicolas PATIN
Compiègne, France
March 2015

# 1

# Introduction to EMC

## 1.1. Problems and definitions

All operational electrical or electronic devices produce interference, which may affect the operation of the device itself and/or that of nearby electrical or electronic equipment. Electromagnetic compatibility (EMC) is a domain which is concerned with the coupling of devices and aims to use all possible means to guarantee the "harmonious" operation of a set of nearby, or coupled, equipment. EMC may be compared to a set of rules for "peaceful coexistence", and is based on a set of standards that must be respected. EMC includes both a scientific aspect, which consists of studying the way in which a device interferes with (or pollutes) its environment, via different types of connections to the "victims", and a standardization aspect, concerning the specification of acceptable thresholds for interference emission, and of sensitivity thresholds at victim level.

The nature of interacting equipment is highly variable. In some cases, elements are truly separate (for example, a television and a telephone); however, they may also form part of the same device (for example, the power supply and motherboard of a personal computer (PC)). Generally speaking, interference may propagate along electric wires (or

PCB tracks): this is known as conducted interference. It may also propagate through empty space (i.e. air or a vacuum) in the case of radiated interference.

"Low-frequency" interference is essentially propagated toward victims by conduction, while higher frequencies are mostly propagated by radiation, as the use of filters allows us to prevent their propagation by conduction. This method is relatively cheap (or natural, given the inductive behavior of connection wires and, for example, the capacitive character of PCB tracks with ground and power supply planes). However, further study is required, as components (inductances and capacitors) are not always able to operate at the frequencies in question.

Conducted and radiated interference will be covered in detail in Chapters 2 and 3; in the case of conducted interference, particular attention will be given to the spectral breakdown of interference (notably for applications connected to the 50 Hz) network:

– electrostatic interference (static electricity, a type of interference often ignored in power electronics);

– very low-frequency interference (flicker, <10 Hz);

– "low-frequency" harmonic interference, of the order of a few multiples of 50 Hz;

– "medium-frequency" interference, linked to the switching frequency (and to its first multiples, for example, from 10 to 100 kHz for industrial speed variation drives);

– high-frequency (HF) interference, linked to the switching time in the switches (>1 MHz);

– environmental interference (cosmic or solar radiation, lightning).

Static interference and very low-frequency interference are specific elements, not directly linked to the switching mechanism; flicker, for example, is linked to variable use (on a human time scale) of electrical energy. "Low-frequency" interference is limited to current switching converters (diodes, thyristors and triacs), and thus also belongs to a specific category.

We will, therefore, focus on the two types of interferences which are most widespread in transistor-based converters (choppers, inverters and switch-mode power supplies) used in forced switching, i.e. "medium and high-frequency" interferences.

Environmental interference and the associated protection equipment will be covered in two specific chapters: one in Chapter 2, in relation to conducted interference, and the other in Chapter 3, concerning radiated interference.

The remainder of this chapter is devoted to sources of interference encountered in power electronics: both "natural" interference (lightning and electrostatic discharge) and artificial interference, created by switching, which is at the heart of the EMC problem for electronic power converters.

## 1.2. "Natural" interference

### 1.2.1. *Static electricity*

When two different materials are rubbed together, static electricity may be produced. This is particularly true in relation to the human body and certain fabrics; as the body behaves in a capacitive manner, discharge may occur on contact with electronic circuits. Fragile and/or poorly protected components may be damaged by this phenomenon, so preventive measures should be taken.

The human body has a surface equivalent to that of a sphere with a diameter of 1 m. Considering the capacitance of a spherical capacitor (with two concentric frames of radius $r_1$

and $r_2$, where $r_1 < r_2$), direct application of the Gauss formula gives us an expression of the capacitance $C$ as follows:

$$C = \frac{4\pi\varepsilon_0\varepsilon_r}{\frac{1}{r_1} - \frac{1}{r_2}} \quad [1.1]$$

In this case, regarding the intrinsic capacitance of the human body, the external frame of radius $r_2$ must be considered to extend to infinity. This gives us the following expression:

$$C_\infty = 4\pi\varepsilon_0\varepsilon_r r_1 \quad [1.2]$$

As we live in air, $\varepsilon_r = 1$, giving a capacitance of 56 pF for $r_1 = 0.5$ m.

This capacitance is clearly affected by proximity to the ground, which adds around 100 pF, and additional capacitances may be added, linked to walls or furniture located close to the body (varying from approximately 50 to 100 pF). This gives an overall parallel association of capacitances of around 200 pF. Note, moreover, that this capacitance is not the only element in the equivalent model of the body when charging or discharging: skin contact is resistive, with a value varying from 500 Ω to 10 kΩ for different individuals and according to the contact surface (the end of a finger or the palm of a hand); this value is also affected by the humidity of the skin. Thus, the human body can be assimilated to a series $R, C$ circuit (an inductance may even be included, with a value of less than 100 nH).

Electrostatic charge can easily reach very high values without the individual in question being aware of it, as voltages under 3.5 kV cannot be felt. Table 1.1 shows two examples of charges produced by walking on two different materials and for two different levels of air humidity.

| Charge generation sources | Humidity: 10 to 20 % | Humidity: 65 to 90 % |
|---|---|---|
| Walking on carpet | 35 kV | 1.5 kV |
| Walking on vinyl flooring | 12 kV | 250 V |

**Table 1.1.** *Examples of electrostatic charge in the human body*

### 1.2.2. *Lightning*

Lightning (see Figure 1.1) is a common natural phenomenon, with an estimated 32 million bolts worldwide each year. Figure 1.2 (a) shows a global map with a color scale representing the number of lightning bolts per square kilometer per year. Note that the majority of lightning occurs close to the equator, and that strikes at sea are considerably rarer (but not unknown). Note, however, that lightning does not simply concern discharge from a cloud to the ground: 60 % of lightning bolts form inside a cloud or between two clouds.

**Figure 1.1.** *Photograph of a storm (source: Wikipedia, Hansueli Krapf)*

France alone receives 1 million lightning bolts per year on an average. The map in terms of numbers of storms per year shown in Figure 1.2(b) shows a high concentration of bolts in mountainous regions, with considerably fewer storms along the coast. The lightning trigger mechanism is primarily based on an increase in the electrical field induced by particular atmospheric conditions (fast-moving masses of air at different temperatures producing significant friction between molecules, generating an electrostatic charge in storm clouds). Once the field reaches a certain threshold, imposed by the dielectric rigidity of the air, a small discharge, known as the "precursor", propagates between a positively charged zone and a negatively charged zone.

These events, generated by strong magnetic fields[1] resulting from the voltages produced (which can reach 100 MV), are extremely short (1/4 s overall), but violent, with pulsed currents of several tens of kA (values of over 200 kA are extremely rare, and the median lies at around 50 kA). This value results in a significant temperature increase, which may reach 30,000°C at ground level; this explains the formation of fulgurites (vitrified sand) at impact points in sandy soil.

Lightning is a dangerous phenomenon, causing the death of 8–15 people and 20,000 animals per year in France. Storms also result in considerable damage due to the intense currents and high temperatures they generate: 20,000 incidents, including 15,000 fires, the destruction of 50,000 electricity meters and of 250 church towers. Clearly, the power involved is sufficient to destroy any electronic equipment without appropriate protection.

### 1.2.3. *Protection equipment*

Protection against lightning is primarily supplied by the network itself, as the peak effect means that the network is the main victim[2]. To avoid the propagation of overvoltage through the network, air spark gaps (see Figure 1.3) are used to limit the voltage and divert the energy from the lightning toward the ground closest to the point of impact.

---

1 The dielectric rigidity of air is 3 MV/m for 11 g/m$^3$ humidity.

2 Pointed objects (such as electricity pylons) increase the electrical field locally, promoting discharge (producing sparks in the surrounding air).

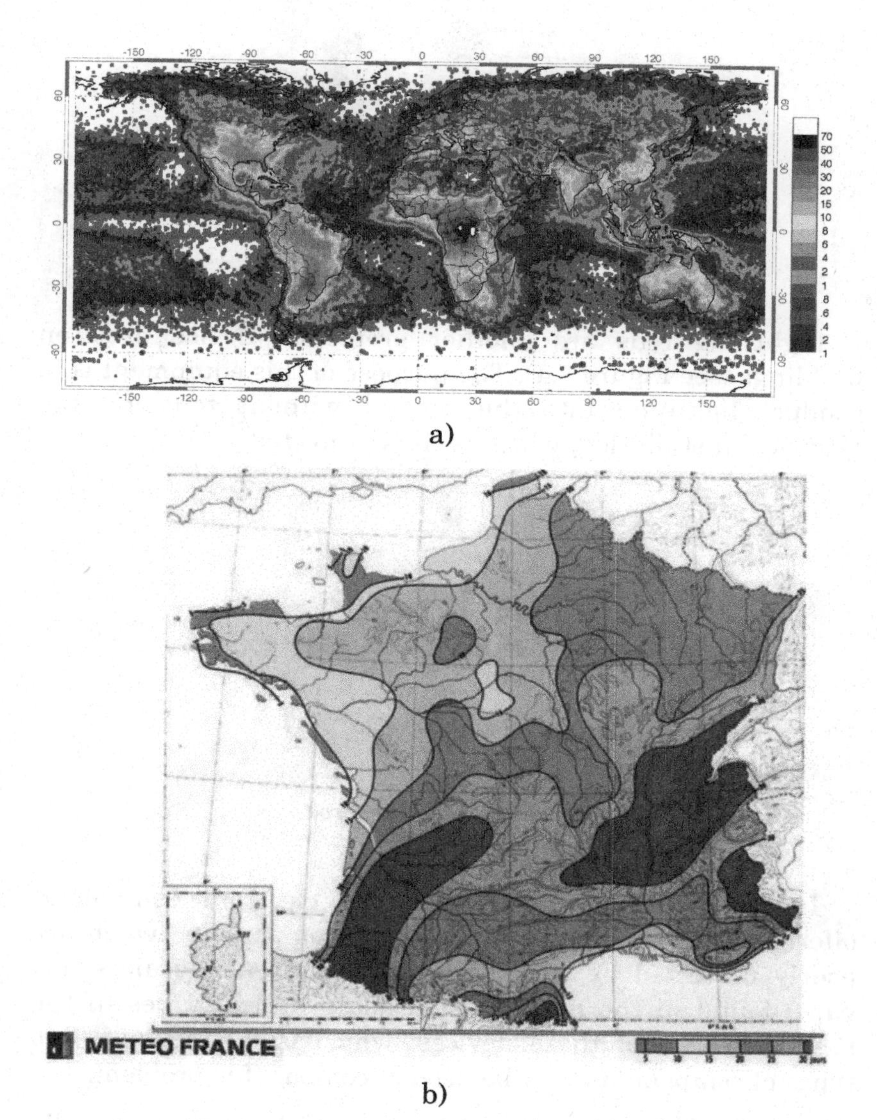


**Figure 1.2.** *Maps showing the occurrence of storms a) worldwide (source: Wikipedia, NASA) and b) in France (source: Météo France / Alain Morel). For a color version of the figure, see www.iste.co.uk / patin / power4.zip*

**Figure 1.3.** *Spark gap*

Lightning rods are another protection device used on buildings (see Figure 1.4). The purpose of this equipment is to conduct lightning into the ground without touching the electrical installation, which is thus protected.

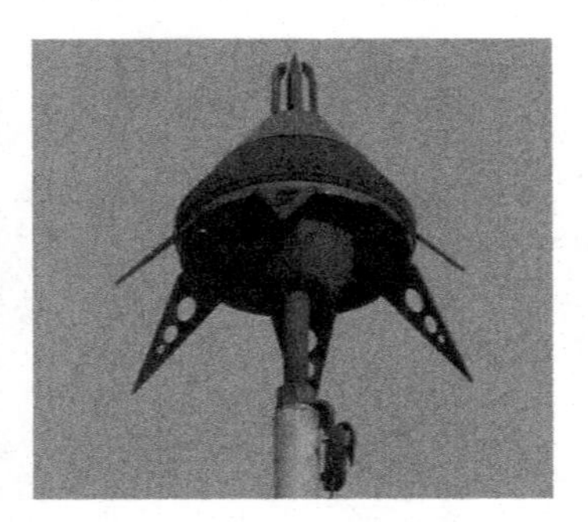

**Figure 1.4.** *Lightning rod*

Individual protection may be used for electronic equipment, offering effective protection of fragile components which are poorly protected by gas spark gaps (particularly air spark gaps) found in the network. The main difficulty lies in the reaction speed of these devices, which is insufficient. Two types of components may be used to combat this problem:

– varistors (nonlinear resistances which decrease rapidly above a certain voltage);

– transil diodes (a specific type of Zener diode).

REMARK 1.1.– All of these components (spark gaps, varistances or transils) should be placed in parallel to the equipment being protected. To guarantee an effective protection, guide UTE C 15-443 recommends cabling of less than 50 cm between the phase to protect and the earth. In practice, care should be taken to connect the power supply terminals directly to the component pins, and then redirect the wires (or tracks on a printed circuit board (PCB)) of the power supply back (toward the protected equipment) from the same terminals. The phenomena which we wish to present are very brief, and it is important to remember that cables have an inductance of around 1 $\mu$H/m; the length of derivation cabling used to connect the protection component should, therefore, be minimized (see Figure 1.5).

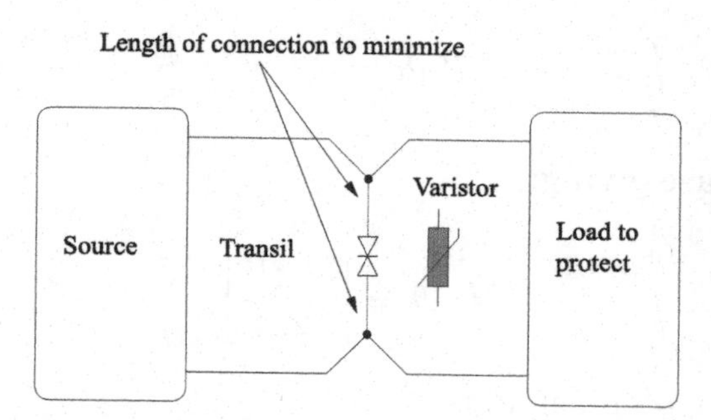


**Figure 1.5.** *Connection recommendations for a transil diode or a varistor*

## 1.3. Switching in power electronics

The switching of high-value voltages and currents can potentially generate significant interference due to the presence of very large $\frac{dv}{dt}$- and $\frac{di}{dt}$-type variations. As an example, for a variable speed drive with a direct current (DC) bus voltage of 600 V and "peak" currents of 28 A in each phase of the powered machine, if the insulated gate bipolar transistors (IGBTs) switch in 400 ns, we obtain $\frac{dv}{dt}$ and $\frac{di}{dt}$ of 1.5 GV/s and 70 MA/s, respectively.

At switching level, the "modulation" aspect of duty ratios may be left aside, leaving only the square waveform (with any given duty ratio $\alpha$); the spectrum of a gate function can therefore be analyzed, with a width considered equal to $\alpha T_d$ in this case[3].

The normalized function $s_0(t)$, defined below, will therefore be studied:

$$s_0(t) = \begin{cases} 1 \, \forall \, |t| \leq \frac{\alpha T_d}{2} \\ 0 \, \forall \, |t| > \frac{\alpha T_d}{2} \end{cases} \qquad [1.3]$$

The spectrum $S_0(f)$ is established using a Fourier transform:

$$S_0(f) \triangleq \int_{\mathbb{R}} s_0(t) . \mathrm{e}^{j2\pi f t} . dt \qquad [1.4]$$

In this case giving:

$$S_0(f) = \int_{-\alpha T_d/2}^{\alpha T_d/2} \mathrm{e}^{j2\pi f t} . dt = \left[ \frac{\mathrm{e}^{j2\pi f t}}{j2\pi f} \right]_{-\alpha T_d/2}^{\alpha T_d/2}$$
$$= \alpha T_d \cdot \mathrm{sin_c}(\pi f \alpha T_d) \qquad [1.5]$$

where the cardinal sine function $\mathrm{sin_c}(\cdot)$ is expressed as:

$$\mathrm{sin_c}(x) \triangleq \frac{\sin x}{x} \qquad [1.6]$$

3 Appendix 2 shows that the spectrum of a $T$-periodic signal is a form sampled at all multiples of the frequency $1/T$ of the continuous spectrum (in the sense of the Fourier transform) of the elementary motif defined for a period.

Given that the sine function is less than or equal to 1 in terms of absolute value, the spectrum of $S_0(f)$ (modulus) can be seen to follow a hyperbolic envelope of the form:

$$\text{Env}\,|S_0(f)| = \frac{1}{\pi f} \qquad [1.7]$$

Representing the spectrum in a log/log reference frame, we obtain a first-order asymptote with a slope of $-20$ dB/dec.

This model of switching in power electronics (which will be known as the zero-order model from now on) is only realistic at low and medium frequencies. In this model, switching is considered to be instantaneous, leaving aside switching times and motifs. A more realistic modeling approach may be obtained by convolving the zero-order model with a motif representing switching. This approach was initially proposed by Costa *et al.* [COS 05, REB 98].

The simplest model of this type (first-order model) consists of convolving $s_0(t)$ by a second port $s_{\text{switch}}(t)$ of width $T_{\text{switch}} \ll T_d$ (representing the switching times, which are considered to be identical for the rising and descending edges of the PWM signal) and amplitude $1/T_{\text{switch}}$ in order to retain the amplitude of $s_0(t)$ after convolution. A signal $s_1(t)$ is thus obtained as follows:

$$s_1(t) = (s_0 \star s_{\text{switch}})(t) \qquad [1.8]$$

This result is presented from a temporal perspective in Figure 1.6. We know (see Appendix 2) that a convolution product becomes a simple product in the frequency domain. Given that $s_{\text{switch}}(t)$ is of the same nature as $s_0(t)$ in the time domain and that the only difference consists of replacing $T_d/2$ (width of the gate function) by $T_{\text{switch}}$, the following result is obtained:

$$S_1(f) = \alpha T_d \cdot \text{sin}_c(\pi f \alpha T_d) \cdot \text{sin}_c(\pi f T_{\text{switch}}) \qquad [1.9]$$

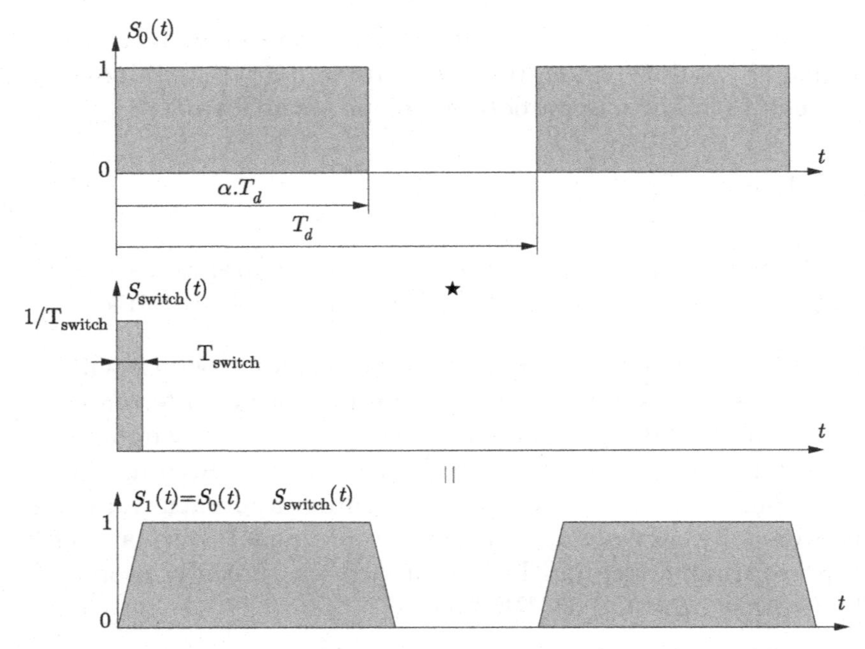


**Figure 1.6.** *Switching spectra (zero-order and first-order models)*

Repeating the envelope calculations gives us:

$$\text{Env}\left[S_1\left(f\right)\right] = \frac{1}{\pi^2 T_{\text{switch}} f^2} \qquad [1.10]$$

The asymptotic behavior of the spectrum thus follows a second-order asymptote (still within the log/log frame) with a slope of −40 dB/dec.

REMARK 1.2.– The envelope is only representative of the HF trend of the evolution of the spectrum in question. The cardinal sine deviates strongly from the hyperbola in the vicinity of zero, in that it has a value of 1 at the origin. In practice, a horizontal asymptote may be added across almost the full width $f_p$ of the main lobe of the narrowest cardinal sine. In this case, the narrowest value of $S_1(f)$ corresponds to $\mathrm{sin_c}\left(\pi f \alpha T_d\right)$, and the equation model gives:

$$\pi f_p \alpha T_d = \pi \qquad [1.11]$$

hence:

$$f_p = \frac{1}{\alpha T_d} \qquad [1.12]$$

Spectra $S_0(f)$ and $S_1(f)$ are represented in Figure 1.7 for $F_d = 10\,\mathrm{kHz}$, $\alpha = 0.5$ and $T_{\mathrm{switch}} = 500\,\mathrm{ns}$.

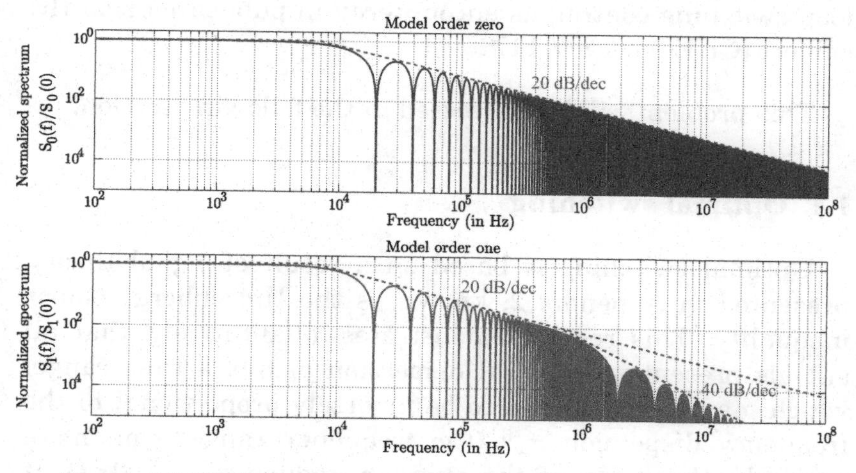


**Figure 1.7.** *Switching spectra (order 0 and order 1 models). For a color version of the figure, see www.iste.co.uk/patin/power4.zip*

The spectrum of the order 1 model (i.e. the closest to the reality of non-instantaneous switching) is less rich in HF than the zero-order model. Presuming that a dynamic of 100 dB constitutes the measurable limit of a spectrum, the order 0 switching model still includes significant components at 100 MHz, whereas there are no measurable components beyond 20 MHz for switching using the order 1 model.

The switching observed in a chopper is in practice nonlinear. The switching profiles present a more complex form, and rising and descending switchings are different. However, it is preferable to retain a symmetrical model, and it is interesting to seek an optimal switching profile for two reasons:

– to establish a minimum limit for this source of interference, which forms the starting point for an EMC study, allowing identification of the most favorable situation for filter dimensioning purposes;

– to propose a switching reference value when using closed-loop switching control, as encountered in publications on the subject [CHE 09, OSW 11][4].

This problem will be addressed in the following section.

## 1.4. Optimal switching

Optimal switching is based on a result of signal theory, presented in Appendix 2, known as the Heisenberg[5]–Gabor inequality. This result demonstrates, qualitatively, that all signals present a temporal dispersion $\sigma_t$ (i.e. a time range) which may be considered to be inversely proportional to the frequency dispersion $\sigma_\omega$[6] (the frequency range – or, more precisely, the range of the angular frequency $\omega = 2\pi f$). It shows that the product $\sigma_t.\sigma_\omega$ is always greater than or equal to $1/2$, irrespective of the signal in question. Moreover, in

4 *A priori* not yet encountered in industrial solutions.

5 Not related to the pseudonym of Walter White (from *Breaking Bad*).

6 Mathematically speaking, this is not strictly true, but it remains satisfactory in qualitative terms.

Appendix 2, equality is shown to be obtained using a Gaussian signal of the form:

$$g(t) = \frac{1}{\sigma_t \sqrt{2\pi}} . e^{-\frac{t^2}{2\sigma_t^2}} \quad [1.13]$$

insofar as the temporal dispersion $\sigma_t$ is clearly used as a parameter in the expression, and its spectrum is also known to be Gaussian, with the form:

$$G(\omega) = e^{-\sigma_t^2 \omega^2} \quad [1.14]$$

In this expression, the angular frequency dispersion $\sigma_\omega$ is seen to be equal to $\frac{1}{2\sigma_t}$, allowing verification of the equality.

This result may be used when studying switching, as $g(t)$ can be used in place of $s_{\text{switch}}(t)$ as a switching motif. This function verifies the only property required for this type of signal:

$$\int_{\mathbb{R}} g(t).dt = 1 \quad [1.15]$$

By convolution with the zero-order model $s_0(t)$, a new switching model is obtained, denoted $s_\infty(t)$ as the Gaussian is infinitely derivable, and its spectrum $S_\infty(f)$ is expressed as:

$$S_\infty(f) = S_0(f) . e^{-\sigma_t^2 \omega^2} \quad [1.16]$$

As the spectrum $S_0(f)$ is imposed from a functional perspective, the weighting introduced by factor $e^{-\sigma_t^2 \omega^2}$ produces a spectrum with the smallest possible spread (i.e. with maximum HF damping), based on the previous result.

A time dispersion equivalent to that of the order 1 model must now be defined in order to carry out a correct comparison. To do this, the following temporal dispersion formula (also given in Appendix 2):

$$\sigma_t = \left( \frac{\int (t-\bar{t})^2 . |\psi(t)|^2 .dt}{\int |\psi(t)|^2 .dt} \right)^{1/2} \quad [1.17]$$

is applied to a "gate" function $p(t)$.

In this case, we therefore have (for a gate of width $T_{\text{switch}}$ centered on the origin):

$$\int |\psi(t)|^2 .dt = \int_{-T_{\text{switch}}/2}^{-T_{\text{switch}}/2} \frac{1}{T_{\text{switch}}^2} dt = \frac{1}{T_{\text{switch}}} \quad [1.18]$$

and:

$$\begin{aligned} \int (t-\bar{t})^2 . |\psi(t)|^2 .dt &= \int_{-T_{\text{switch}}/2}^{-T_{\text{switch}}/2} \frac{t^2}{T_{\text{switch}}^2} dt \\ &= \frac{1}{T_{\text{switch}}^2} \left[ \frac{t^3}{3} \right]_{-T_{\text{switch}}/2}^{T_{\text{switch}}/2} \\ &= \frac{T_{\text{switch}}}{12} \end{aligned} \quad [1.19]$$

Finally, we obtain:

$$\sigma_t = \frac{T_{\text{switch}}}{\sqrt{12}} \simeq 0,289.T_{\text{switch}} \quad [1.20]$$

or $T_{\text{switch}} = 3.464.\sigma_t$.

For illustrative purposes, a "gate" function and a Gaussian with the same temporal dispersion are shown superimposedly in Figure 1.8. Most of the Gaussian is located within the temporal support of the gate (i.e. outside of the interval $[-T_{\text{switch}}/2; T_{\text{switch}}/2]$). By integrating the Gaussian numerically between $[-1.732.T_{\text{switch}}; 1.732.T_{\text{switch}}]$, we obtain:

$$\int_{-1.732}^{1.732} \frac{1}{\sqrt{2\pi}} e^{-\frac{t^2}{2}}.dt \simeq 0.917 \qquad [1.21]$$

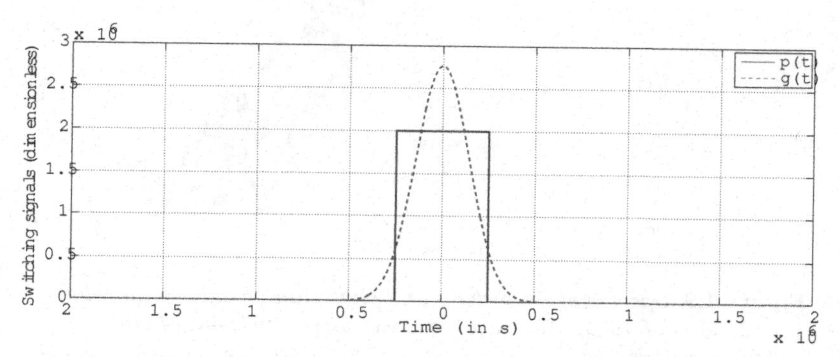


**Figure 1.8.** *Switching motifs $s_{\text{switch}}(t)$ of the "door" and "Gaussian" types. For a color version of the figure, see www.iste.co.uk/patin/power4.zip*

Thus, this result can be practically assimilated to a switching operation measured (experimentally or by simulation) between 5 and 95 % of the variation of the switched signal. This result is useful for evaluating the performance of switching obtained based on experimental results, or for more or less fine simulation of real components (for example, using a SPICE-type simulator).

Having established the relevance of this type of "Gaussian" switching motif in terms of EMC, we need to produce the corresponding spectrum graph as shown in Figure 1.9. For illustrative purposes, the figure also shows the trace of the first-order model. In equivalent conditions, the measurable limit frequency (vertical dynamic fixed at 100 dB for a first-order model) is around 20 MHz, whereas for the Gaussian model the limit is between 2 and 3 MHz. This switching profile, therefore, produces significant gains.

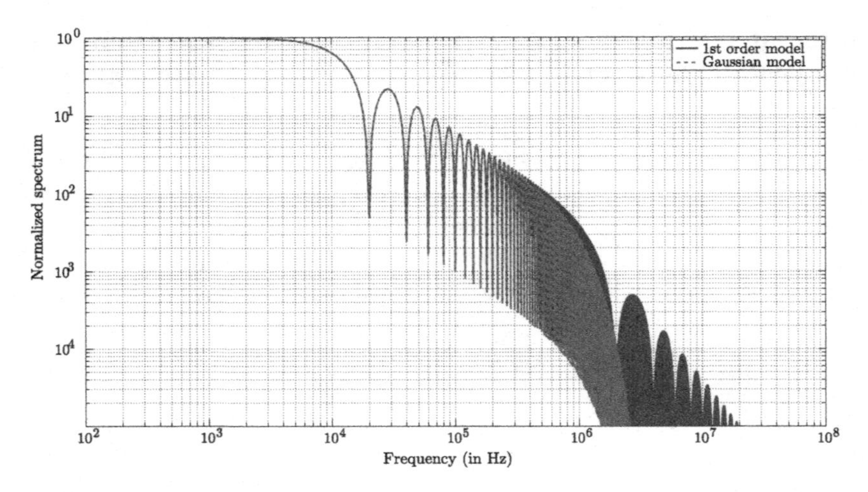


**Figure 1.9.** *Spectra of the order 1 and Gaussian models. For a color version of the figure, see www.iste.co.uk/patin/power4.zip*

The graph of the Gaussian model, therefore, constitutes an absolute minimum, only parameterized by $\sigma_t$, i.e. a switching time $T_{\text{switch}}$ fixed by constraints external to EMC (switching losses, thermal losses, physical limits of the switching component(s), etc.). This boundary may be seen either as an unattainable theoretical limit or as an objective to aim for when designing advanced gate drive circuitry for a metal oxide semiconductor field effect transistor (MOSFET) or IGBT.

## 1.5. Standardization

EMC is not only a science (more specifically, a branch of electronics, in the broadest sense), but also the "art" of establishing successful coexistence between devices. To do this, rules are needed to define the "rights" and "responsibilities" of each electrical and/or electronic device so as to guarantee correct operation. This definition was presented at the beginning of this chapter, but should be retained in this context: EMC cannot be limited to the simple analysis of interference phenomena, but it should also limit their impact. The evaluation framework is therefore arbitrary, and is defined by means of standards. A variety of standards exist according to:

– the domain of application of the equipment in question: industrial, medical, domestic, automobile, rail, naval, aeronautics, civil or military;

– the nature of the interference source: conducted artificial interference, radiated artificial interference, static electricity, lightning, cosmic radiation, etc.

A considerable amount of information concerning standards is available in [IEE 12], bringing together all of these categories via current and past standards (from the 1950s to the present). The collection notably includes the standards defined by the International Electrotechnical Commission (IEC 61000 standard) and refined by the *Comité International Spécial des Perturbations Radioélectriques* (CISPR, special international committee on radio interference), founded in 1934[7], which has published a certain number of documents:

– CISPR 10: organization, rules and procedures of the CISPR;

– CISPR 11: industrial, scientific and medical (ISM) equipment – radio-frequency disturbance characteristics – limits and methods of measurement;

– CISPR 12: vehicles, boats and internal combustion engine driven devices – radio disturbance characteristics – limits and methods of measurement for the protection of receivers themselves, except for those installed in vehicles/boats/engines themselves or in nearby vehicles/boats/engines;

7 Proof that these are not simply recent concerns.

– CISPR 13: radio and television receivers and associated equipment – limits and methods of measurement;

– CISPR 14-x: electromagnetic compatibility – requirements for household appliances, electric tools and similar apparatus – in two parts: emissions and immunity;

– CISPR 15: limits and methods of measurement of radio disturbance characteristics of electrical lighting and similar equipment;

– CISPR 16: specification for radio disturbance and immunity measurement apparatus and methods;

– CISPR 20: radio and television receivers and associated equipment – immunity characteristics – limits and methods of measurement;

– CISPR 22: information technology equipment – radio disturbance characteristics – limits and methods of measurement;

– CISPR 24: information technology equipment – immunity characteristics – limits and methods of measurement.

In the case of aeronautics, onboard equipment is subject to Joint Aviation Requirements (JAR) certification, for example, which includes an "EMC" component.

## 1.6. Summary

In this introductory chapter, a certain number of definitions and terms used in EMC have been established. A significant term for interference sources has also been highlighted, in the form of the switched quantity in a converter controlled by PWM. This approach is fundamentally oriented toward transistor-based converters in forced switching; this restriction is, in reality, suitable for use with the majority of the modern electronic power converters which have been analyzed elsewhere in this series. The source term will be used as an input quantity in filter dimensioning problems in order to comply with the acceptable interference ranges established by the relevant standards. The "filtering" aspect of EMC will be considered in the following chapter, along with "circuit" coupling mechanisms (i.e. with lumped elements) between interference-emitting equipment and the victim.

# 2

# Lumped Parameter Models

## 2.1. Context

Conducted interference represents a significant part of the interference covered in electromagnetic compatibility (EMC). As the name indicates, this interference uses wires as a support for propagation from a source to a victim. Both common impedance couplings, which are evidently responsible for conducted disturbances, will be covered in this chapter; we will also consider interference through inductive or capacitive coupling which, in a certain way, are associated with radiation of the magnetic or electrical field, respectively. In this case, all forms of couplings which may be modeled using an electrical diagram with lumped elements will be considered to be conducted disturbances. The fundamental difference between coupling types lies in the frequency of the electrical quantities (voltage and/or current) involved, along with the dimensions (in the geometric sense of the term) of the device in question.

In the case of a circuit that is small compared to the associated wavelength ($\lambda = c/f$ in a vacuum[1], slightly lower in dielectrics such as FR4 or Teflon), all couplings may be

1 Where $c$ is the speed of light, i.e. $3 \times 10^8$ m/s (see Chapter 3).

modeled as parasitic capacitances and mutual inductances. In these conditions, an EMC study involves the calculation of transfer functions, and may use the layering principle to analyze the contributions of different sources of interference for a victim. This type of modeling and analysis will be discussed in this chapter. In some ways, this type of study is simpler than the study of radiation over significant lengths of the wavelength as, in practice, above $0.1\lambda$, the lumped element approach ceases to be suitable and should be replaced by a common constant approach (based on partially derived equations).

## 2.2. Common impedance interference

Common impedance interference is extremely widespread in the 50 Hz electric network (with, ideally, a sinusoidal voltage in open circuit mode), distributed across a variety of equipment, absorbing currents that are often non-sinusoidal. If a source is connected to these different loads via a common impedance $Z_{\text{switch}}$, a voltage drop occurs at the impedance terminals, and this will be experienced by all of the loads placed in parallel.

In practice, this phenomenon only becomes visible when a powerful nonlinear load is placed on the network:

– a high-power installation (industrial or otherwise: e.g. a reflow furnace, or, on a smaller scale, a welding machine);

– multiple synchronized low-power loads (such as capacitance-head rectifiers).

A typical configuration is shown in the diagram of Figure 2.1.

Based on the diagram, the expression of the voltage $V_{load}$ at the terminals of the two loads can be written as:

$$V_{load} = V_{net} - Z_{\text{switch}} \cdot (I_{load1} + I_{load2}) \qquad [2.1]$$

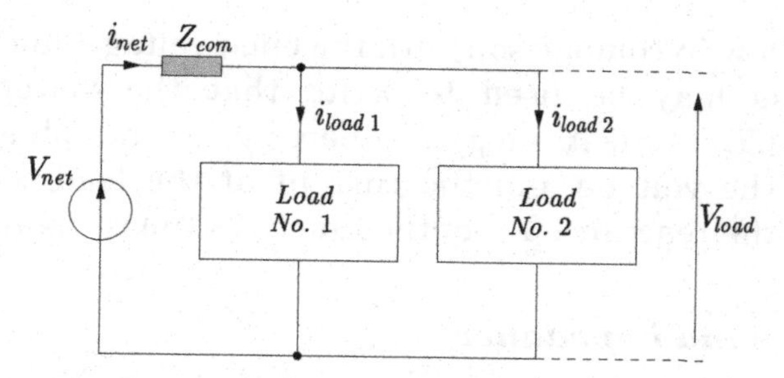


**Figure 2.1.** *Common impedance coupling*

Consequently, load 1 "sees" not only the voltage drop $Z_{\text{switch}}.I_{load1}$ resulting from its own consumption, but also a voltage drop $Z_{\text{switch}}.I_{load2}$ caused by load 2. This drop can be significant in the case of "powerful" loads. Note, moreover, that this voltage drop is not necessarily resistive, but also inductive in distribution networks.

### 2.2.1. *Flicker*

EMC interference is not only caused by power switching, where electronic components are used in switching mode; it may also be caused by slow fluctuations in the power (active and reactive) consumed by a load. As we have seen, this fluctuation generates a variation in the power voltage observed by the fluctuating load itself, but also by other parallel loads (particularly lighting). The low-frequency voltage fluctuation phenomenon[2] is particularly troublesome for human beings (leading to fatigue, irritability and epileptic episodes) in the context of lighting variations. The flicker phenomenon should be limited in order to correspond to strict standards (IEC 61000).

Generally speaking, voltage drops (both resistive and inductive) in the network (modeled by a series impedance

2 With periods much longer than the network period, and than the time constant of the human eye (retinal persistence).

$R, L$) are low in comparison with the open voltage, and Kapp's hypothesis may be used to write that the voltage drop $\Delta V = (R.\cos\varphi + X.\sin\varphi).I$ where $\varphi$ is the phase shift between the voltage and the current at the load terminals and $X$ is the reactance $L\omega$ of the line ($R$ is the resistance).

### 2.2.2. *Ground impedance*

The common impedance coupling phenomenon is observable for a wide range of frequencies and for all power supply types (from alternating current (AC) or direct current (DC) networks to printed circuit boards (PCBs)[3]). A typical example of common impedance coupling occurs in "badly designed" PCBs. In a diagram, the ground is represented by a symbol showing the reference potential (conventionally 0 V), presumed to be common to all points of the connected electronic assembly. Physically speaking, however, interconnected copper tracks present a resistance (connected to the cross-section, i.e. the width and thickness of the track, the length and the current frequency, generating a skin effect). These conductors also present an inductance linked to the track length and geometry (and, potentially, to the environment, if ferromagnetic materials are present). In these conditions, an impedance of the form $R + jL\omega$ occurs, such as that seen in the previous section. There may therefore be a potential difference between two ground points if a current is circulating (particularly if the current has a high variation frequency) between these points. This situation may lead to dysfunctions, and can also cause the destruction of certain sensitive components on a PCB subject to these fluctuations in "ground" potential.

3 A PCB is an isolating plate (generally made up of epoxy resin and fiber, but which may also be made from Teflon for certain high-frequency (HF) applications), with thin copper connection tracks (35 $\mu$m is standard), used for integrated circuits (and/or discrete components) which are soldered in place (either using holes and pins or directly onto the surface of the PCB).

In these conditions, we must ensure that:

– the track impedance is sufficiently low (short tracks, sufficiently wide and thick[4]);

– the ground equipotentials (this may also apply to other equipotentials) should, as far as possible, be connected to a common point (constituting a star connection).

Generally, components should be placed onto a PCB in a way which simplifies the connections as far as possible (i.e. shortening connections and avoiding diversions). This task may be made easier using modern computer-aided design (CAD) tools, but it is often advisable to group components by "function" in order to simplify manual routing.

The use of ground planes (large areas of copper, such as that shown in Figure 2.2) reduces the ground impedance, and fulfills other functions which are useful for EMC (screening, to avoid cross talk between tracks with different signals – see the next section).

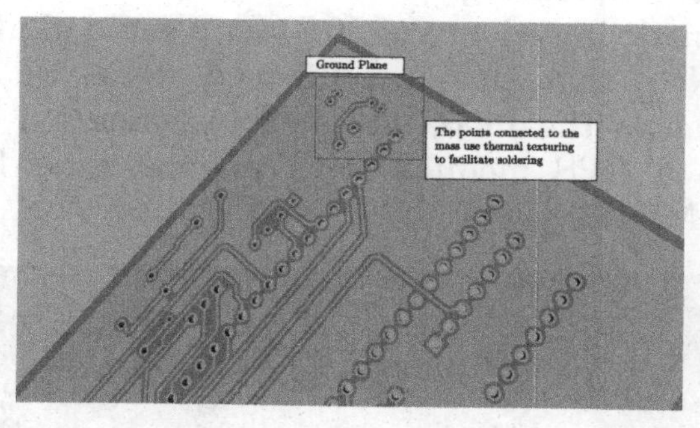


**Figure 2.2.** *PCB with a ground plane. For a color version of the figure, see www.iste.co.uk/patin/power4.zip*

4 Width and thickness are generally chosen as a function of the current that needs to circulate (tables are available, giving values based on tolerated self-heating levels – see Figure 6.1 of Chapter 6 in Volume 1).

Finally, the use of multilayer technology (double-sided, in the simplest case, but large numbers of internal layers may be used in modern PCBs) increases the number of degrees of freedom and simplifies routing in complex chips. In RF applications, the inductive character of the metallic vias used for the interconnection of layers should also be taken into account. A via with a length of 1 mm presents an inductance of around 1 nH. If this value is too high, a possible solution (as shown in Figure 2.3) is to multiply the number of vias (placed in parallel) to reduce the equivalent inductance: these are known as *stitching vias*. The figure shows that the vias (in green) are directly connected to a copper plate, shown in red (TOP layer of a double-sided circuit) and appear to be connected in the same way to a second copper plate, shown in blue (BOTTOM layer of the circuit). These two planes act as the ground of the circuit. In the center of the PCB, we also see a copper track (TOP side) surrounded by vias, fulfilling a blinding function (in the same way as a coaxial cable) of the track; the shape and width of this track suggest a controlled impedance operating mode (e.g. microstrip, with a characteristic impedance of 50 Ω) for a rapid digital or analog signal. Note that this is confirmed when we look at the application of the PCB, designed to connect a "patch"-type aerial to a GPS module (with an operating frequency of 1,575.42 MHz).

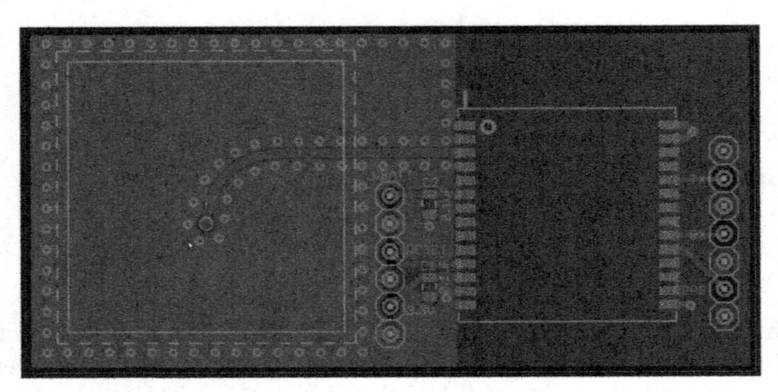

**Figure 2.3.** *Stitching vias between two ground planes. For a color version of the figure, see www.iste.co.uk/patin/power4.zip*

Unfortunately, a copper track cannot be treated in isolation from the other tracks on a PCB. Even if two tracks are completely independent of operational terms when designing an electronic assembly, physical coupling may still occur as a result of proximity.

## 2.3. Coupling interference

In this section, we will show that two nearby tracks can interact. In these conditions, a “residue” of the signal carried by one track may be observed in the second track: this is known as cross talk. This type of phenomenon is particularly critical when a low-amplitude signal is carried by a track (or wire) running parallel to a track (or wire) carrying a high current (or high voltage) with HF variation. Two physical mechanisms are involved in these couplings:

– inductive-type coupling (by mutual inductance);

– capacitive-type coupling.

### 2.3.1. *Inductive coupling*

In inductive coupling, two nearby tracks may be assimilated to the primary and secondary of a transformer. This means that in addition to their individual inductances, the two tracks are connected by a mutual inductance, resulting in the equation system seen in Chapter 5 of Volume 1 [PAT 15a]:

$$\begin{cases} \psi_1 = L_1.i_1 + M_{12}.i_2 \\ \psi_2 = M_{21}.i_1 + L_2.i_2 \end{cases} \qquad [2.2]$$

where $L_1$ and $L_2$ are the separate inductances of each track and $M_{12} = M_{21} = M$, given the mutual reciprocity of the actions of the two tracks. Fluxes $\psi_1$ and $\psi_2$ are connected to the potential differences at the terminals of the two tracks (see Figure 2.4) as follows:

$$\begin{cases} \Delta v_1 = R_1.i_1 + L_1\frac{di_1}{dt} + M\frac{di_2}{dt} \\ \Delta v_2 = R_2.i_2 + L_2\frac{di_2}{dt} + M\frac{di_1}{dt} \end{cases} \quad [2.3]$$

This equation model is simple, but the separate and common inductances in the two tracks are not easy to calculate. Analytical formulas may exist, but in this case, they are often applied to complex geometries (turns, vias, etc.) and it is better to use specific software in order to obtain satisfactory results. The best-known CAD platforms may include tools to analyze signal integrity.

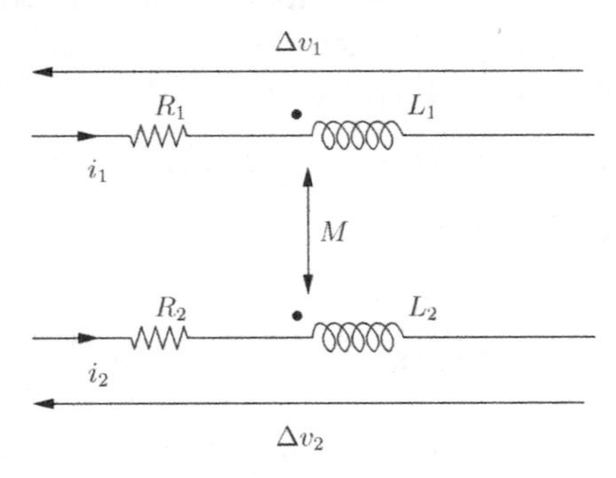


**Figure 2.4.** *Inductive coupling between nearby lines*

A simple rule to minimize track inductances (of the order of 1 nH/mm) and mutual inductances is to reduce track length, as, for example, between a decoupling capacitor and the component subject to (power) voltage smoothing. In this case, the use of vias to and from the ground (GND) and power (POWER) layers is particularly useful, as shown in Figure 2.5 (left). When decoupling needs to operate at very high frequencies, capacitors may even be placed on the other side of the PCB – see Figure 2.5 (right) – to the associated component (in this case, a U2 in a ball grid array (BGA) packaging). More detailed information on the decoupling issue for fast digital circuits is given in the Xilinx technical documentation [XAP 05], notably with regard to the required size of decoupling capacitors (surface mount device (SMD) packages), the technology used and their location on the PCB. Note that this information is also applicable in the context of power electronics.

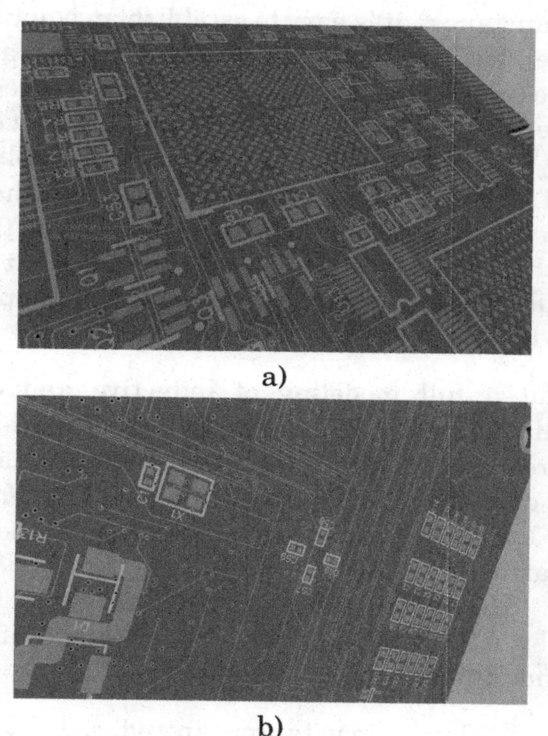


**Figure 2.5.** *Minimization of track length – in this case, for decoupling condensers: a) TOP layer and b) BOTTOM layer. For a color version of the figure, see www.iste.co.uk/patin/power4.zip*

### 2.3.2. *Capacitive coupling*

In addition to inductive coupling, capacitive coupling may occur between nearby lines, as all of the components of a capacitor are present:

– two conductors, placed face-to-face;

– an insulating gap between the two conductors (not only air, but also dielectric substrate, with a permeability going from 4, for FR4, to around 10n for the Teflon used in radio frequency (RF) applications).

As for inductances, it is easy to establish an equation model (circuit type), but the effective calculation of the capacitance between two conductors can be tricky. Note, moreover, that the interference transmitted by capacitive coupling is a current of the form $C\frac{dV}{dt}$, whereas in the case of inductive coupling, it was a voltage of type $M\frac{di}{dt}$. A capacitive coupling will therefore be sensitive to high variations in voltage, notably at the output point of a chopper or inverter (i.e. at the mid-point of a half-bridge), while inductive coupling is sensitive to chopped input currents in these converters.

Note that a full modeling of inductive and capacitive couplings may be obtained in an analytical or semi-analytical way for wired or mass conductors using the partial element equivalent circuit (PEEC) method, notably implemented in CEDRAT's InCa3D program. Details of this method may be found in articles [ROU 04a, ROU 04b] in *Techniques de l'Ingénieur*. Other programs, such as Q3D Extractor, developed by ANSYS, are based on the method of moments [GIB 07] and fulfill a similar role.

REMARK 2.1.– The capacitances introduced by coupling between conductors located opposite one another (tracks, cables, bus bars, etc.) is often negligible in comparison with the capacitances introduced physically by capacitors, particularly for decoupling power supplies. However, their role may still be significant at high frequencies, where electrolytic capacitors cease to function.

To minimize capacitive coupling between tracks on a PCB, a rotation of 90° is carried out between layers (red and blue tracks) during routing, as shown in Figure 2.6.

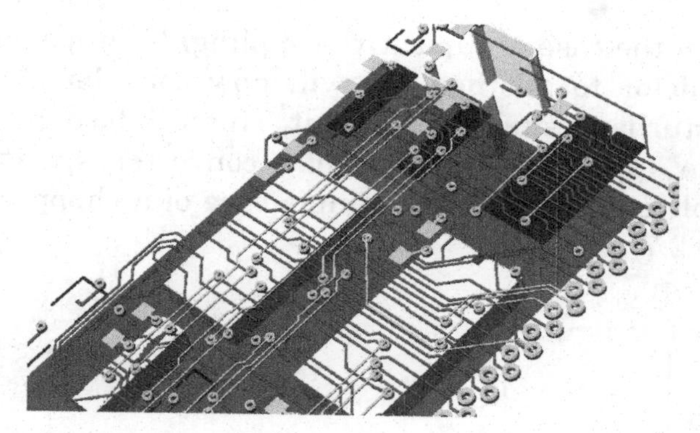

**Figure 2.6.** *Coupling reduction in tracks on two layers of a PCB. For a color version of the figure, see www.iste.co.uk/patin/power4.zip*

There are two ways of minimizing coupling between two parallel signal tracks on a given layer of a PCB:

– increasing the distance between the tracks to reduce coupling (capacitive and inductive)[5];

– place a ground screen between the tracks (a ground plane, or a track connected to the ground at both ends).

This latter solution operates by screening, and is not only applicable to printed circuits, but also to multiwired "ribbon" connection carrying rapid signals (which are often digital)[6].

## 2.4. Interference modes

The previous section covered the physical means used by interference (voltage, in the case of inductive coupling, or

5 This is often difficult when producing a compact PCB, something which also reduces the length of the tracks (and consequently the individual inductances of the connections).

6 In multiwired connections, differential signals are also (and increasingly) used: in this case, two adjacent wires operate as a two-wire line (HDMI, USB, etc.) and offer very good immunity to common-mode interference (a definition of common mode is provided in the following section).

current, in the case of capacitive coupling) to propagate from one conductor to another. We will now consider the paths which disturbances take in a circuit. This will be done using a "multipole"-type representation of a converter, specifically a quadrupole converter, in the simple case of a chopper shown in Figure 2.7[7].

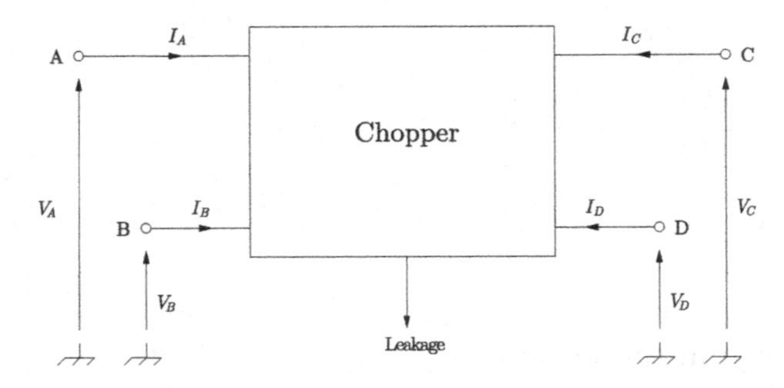


**Figure 2.7.** *Quadrupole (+1 leakage terminal) representation of a chopper*

The diagram includes two input terminals (i.e. the chopper power supply). Generally speaking, we presume that:

– the terminals are at potentials $V_A$ and $V_B$ which are different from 0 V (i.e. the ground);

– they carry different currents $I_A$ and $I_B$ (a leakage pathway is therefore presumed to exist).

The general equation model of the two voltages (and the two currents) includes the introduction of terms which are said to be either:

– differential mode (the difference, up to a coefficient);

– common mode, denoted as $cm$ (the average).

7 T, strictly speaking, is a quadrupole, as there is a fifth (leakage) terminal in this representation of the chopper.

In this case, we note:

$$\begin{cases} V_d = V_A - V_B \\ V_{cm} = \frac{V_A + V_B}{2} \\ I_d = \frac{I_A - I_B}{2} \\ I_{cm} = I_A + I_B \end{cases} \quad [2.4]$$

REMARK 2.2.– Note the difference between the definitions of common and differential mode for voltages and currents. A coefficient of 1/2 may be applied to either the common or the differential mode as a function of requirements; this has a direct effect on the equivalent diagram representation, as we will see.

The decoupling of the two modes allows a two-stage modeling process to be used for the converter and its environment, facilitating the study of conducted interference within the system.

### 2.4.1. *Differential mode*

Differential mode consists of observing the potential difference between the two terminals, considering the two terminals of the multipole to be isolated (i.e. operating as a dipole), and considering a current entering the system through one terminal and leaving, in its entirety, through the other terminal. In the case of input into a chopper or inverter-type converter, the connected source must be of the voltage type, as the converter behaves as a current source. The converter is therefore modeled in the form of an equivalent current source $I_d$. Conversely, the output from a chopper behaves as a voltage source $V_d$ in the context of differential mode. The equivalent model of this converter is shown in Figure 2.8.

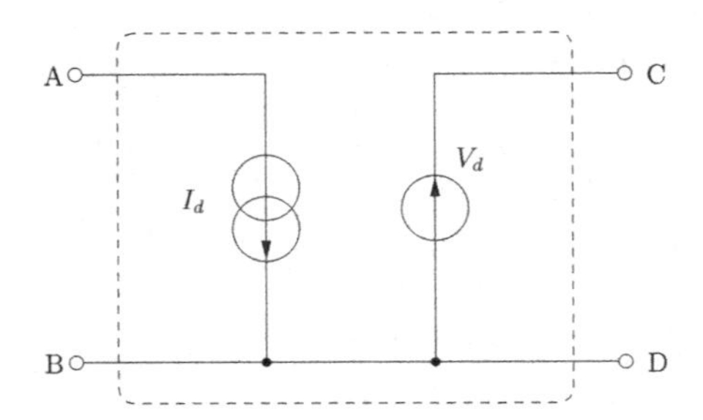


**Figure 2.8.** *Idealized differential model of a chopper*

REMARK 2.3.– The leakage connection is not shown in the differential model.

### 2.4.2. *Common mode and hexapolar representation*

Rather than using a quadripolar structure with an added leakage terminal (see Figure 2.7), it is generally better to use a representation with an even number of terminals, in this case, a hexapole, which may be modeled in matrix form (as in the case of quadrupoles). There is therefore a common terminal for the input and output ports (three terminals per port). This approach is the same as that used for the quadrupole representation with $h$ parameters of the transistor which, while only including three physical terminals, is traditionally represented by a four-terminal model. The equivalent hexapolar model of a chopper is shown in Figure 2.9.

The figure shows the same sources present in the differential mode representation, both on the input and output sides of the converter. Impedances are then added, connecting the different (useful) terminals of the converter to the heat sink: these paths may all be used by leakage currents, generating common-mode interference. These impedances, denoted as $Z_{CC}$, $Z_{EE}$ and $Z_{CE}$, include a series capacitance due to the isolation of the silicon chip in the power component from the metal base which allows cooling.

However, these impedances are not, strictly speaking, purely capacitive due to the connections (bonding wires and possibly printed tracks) between chips inside a module (e.g. an insulated gate bipolar transistor (IGBT) half-bridge)[8]. The "parasite component" aspect of components used in power electronics will be discussed further in the following section.

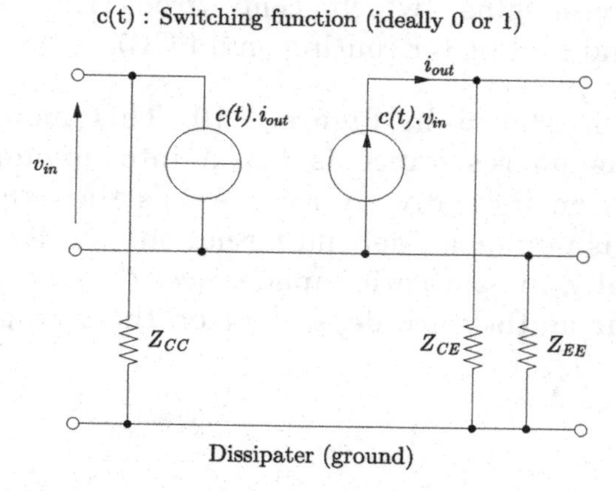


**Figure 2.9.** *Full idealized model of a chopper*

### 2.4.3. *Parasite components in switches*

The chopper models (in differential and common mode) presented above are idealized, which only include current and voltage sources. In reality, the switches making up a half-bridge present parasitic capacitive and inductive behaviors, which must be taken into account in order to show the fastest transitional phenomena which occur during switching (i.e. the HF components of the observed quantity spectra).

8 This is also true in the case of discrete-component converters mounted on PCBs, with the inductances of the component packaging themselves, and the inductances of the tracks in the printed circuit.

#### 2.4.3.1. *MOSFET transistors*

A MOSFET transistor behaves as a current source $I_{DS}$ (between a drain and a source) controlled by the voltage $V_{GS}$ applied between the gate and the source. However, the capacitances present between the three pins must be taken into account, along with the inductances involved in accessing these pins, which come from the component packaging and cabling (or routing on a PCB).

A model is shown in Figure 2.10. This model is still simplified, as no resistance is taken into account for pin access; however, it is already complex, as the expression of current $I_{DS}$ is nonlinear, depending not only on $V_{GS}$, but also on $V_{DS}$. Finally, the parasitic capacitances $C_{GS}$, $C_{DG}$ and $C_{DS}$ are nonlinear as they are dependent on the voltage at their terminals.

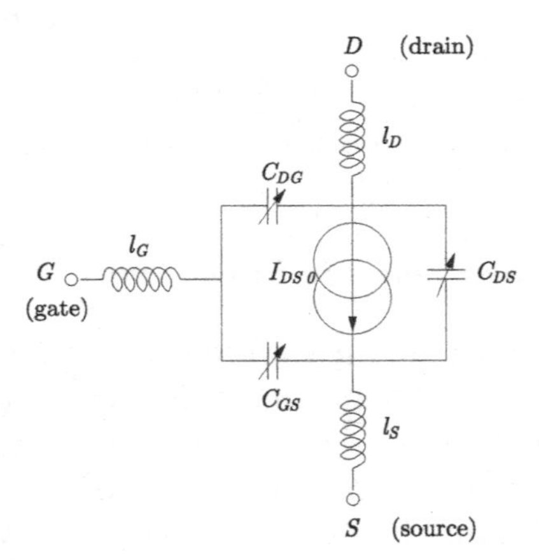


**Figure 2.10.** *EMC model of a MOSFET transistor*

REMARK 2.4.– This diagram essentially corresponds to those used in SPICE models, as supplied by manufacturers, with the exception of the inductances, which are generally linked to the cabling carried out by the component user.

#### 2.4.3.2. *Diodes*

Diodes behave in a nonlinear manner, with two distinct representations in ON and OFF state (see Figure 2.11).

Ideally, the diode is considered to be a voltage source $V_F$ when switched on, and an open circuit when switched off. For a more realistic representation, the dynamic behavior of the diode also needs to be taken into account:

– in the ON state, a dynamic resistance $r_D$ is present, associated with a diffusion capacitance $C_D$;

– in the OFF state, a transition capacitance $C_T$ (nonlinear, as it is dependent on the applied voltage) is present, potentially associated with a resistance (to take account of the leakage current).

#### 2.4.3.3. *"Half-bridge" switching cells*

Complex models of transistors and diodes may make them hard to use in system simulations. It is therefore better to use a simplified model, based on the idealized structure, including the differential and common-mode behaviors as seen in Figure 2.9. Figure 2.12 clearly shows the parasite impedances through which common-mode currents are able to circulate toward the metal base.

## 2.5. Modeling the converter environment

We do not intend to present an exhaustive list of components and the associated models in terms of EMC. In this section, we will focus on examples which are representative of the different components of a switch-mode power supply or a variable speed drive. These two examples will be covered in the next section.

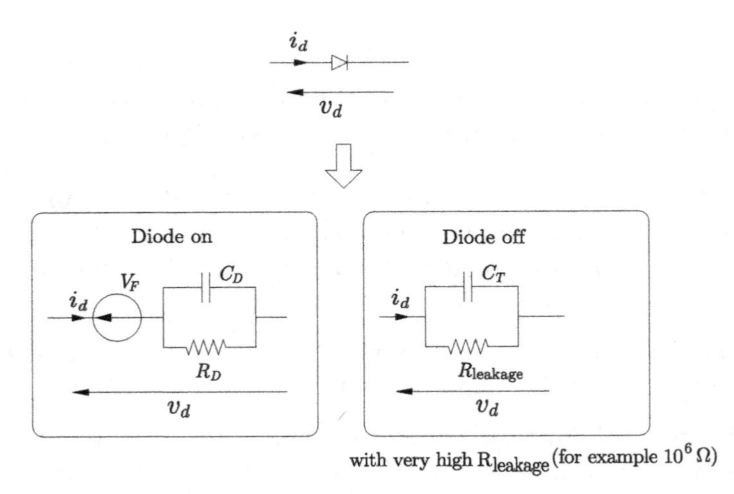


**Figure 2.11.** *EMC model of a diode*

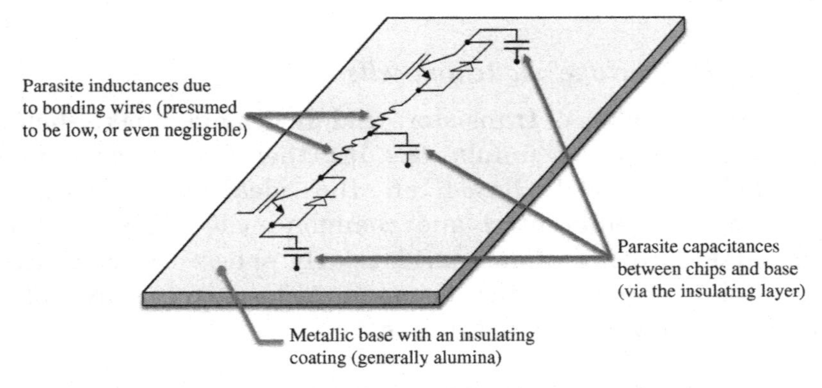


**Figure 2.12.** *"Half-bridge" model and parasite elements*

### 2.5.1. *Modeling cables*

When modeling cables, a certain number of impedances linked to underlying physical phenomena must be evaluated:

– the resistance and inductance of wires;

– the capacitance and conductance of the insulating material between the wires.

Figure 2.13 shows an equivalent model of a shielded three-phase cable, including:

– the impedance, presumed to be zero, of the shielding, assimilated to an ideal conductor;

– an absence of mutual inductance between the wires and the shielding.

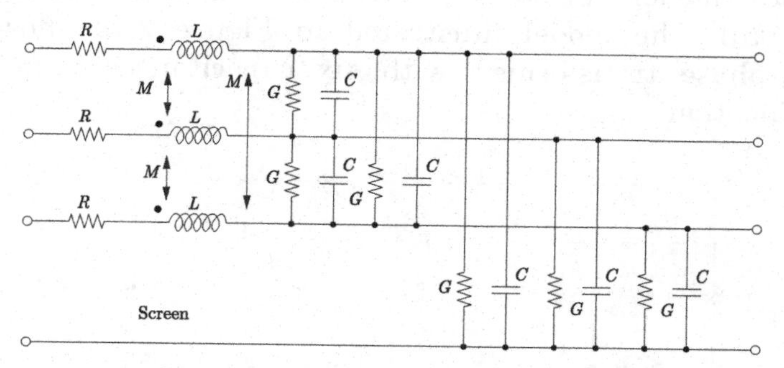


**Figure 2.13.** *Common-mode filter in a USB cable*

REMARK 2.5.– The proposed model is a lumped element model, representing the operation of the cable for frequencies below a certain threshold. As we have already seen, and as we will see in more detail in the next chapter, if the length of the cable is longer that a given fraction of the wavelength $\lambda = \frac{c}{f}$, or more generally $\frac{v}{f}$, associated with the carried signal[9] (in practice, a limit of $\frac{\lambda}{10}$) is used, then this representation is no longer satisfactory, unless a cascade of cable "slices" is considered, each with a model of the type shown in Figure 2.13. In this case, the linear parameters of the cable need to be taken into consideration, and the representation takes the form of a partially derived equation known as the wave equation.

9 Where $v$ is the speed of wave propagation in the dielectric (less than or equal to $c$).

### 2.5.2. *Modeling transformers*

The transformer model is based on the diagram shown in Chapter 5 of Volume 1 with the addition of parasite capacitors in the primary and secondary. The number of capacitors varies as a function of the desired accuracy of the model, particularly at high frequencies. A variety of more less complex models exist (e.g. three or six capacitances in [BRÉ 05]). The model illustrated in Figure 2.14 shows a single-phase transformer, with six capacitances requiring identification.

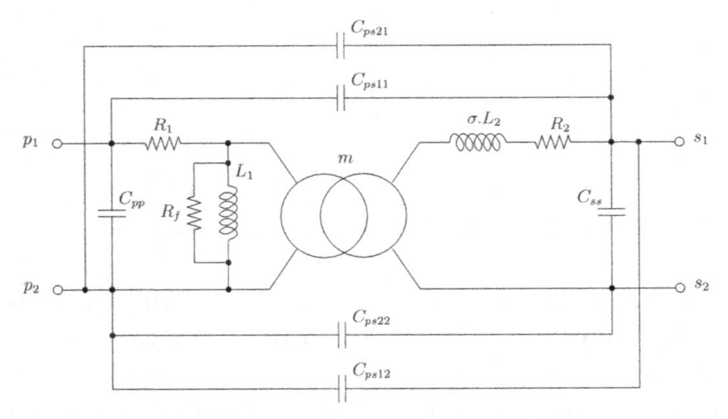


**Figure 2.14.** *Model of a single-phase transformer at HF*

This type of model (which presents a high number of degrees of freedom) demonstrates good theoretical tracking of the real transformer; however, this increased complexity requires considerably more measurements to be taken for parameter identification purposes (and the sensitivity of parameters in relation to measurements needs to be suitable for a valid model to be produced).

REMARK 2.6.– This type of model can potentially be applied to any type of transformer (50 Hz or switch-mode power supplies, or even for transistor gate control – pulse transformers).

### 2.5.3. *Modeling three-phase motors*

The final element which we will model here is the electrical machine; in this case, we will consider the classic case of a squirrel-cage induction motor (the form most widely used in industry). This model can easily be transposed for synchronous motors (and DC motors) as the modeling strategy is similar to an EMC perspective; in EMC, the low-frequency domain, where electromechanical energy conversion occurs, is not taken into account. We focus instead on the HF domain, where coils present inductive and capacitive behaviors which are relatively independent of the mechanical parameters of the machine (at least in the case of machines with smoothed poles). The equivalent electrical diagram of an induction motor from a "low-frequency" perspective (see Figure 2.15) clearly shows physical phenomena such as the magnetization of the magnetic circuit, with a magnetizing inductance $L_m$.

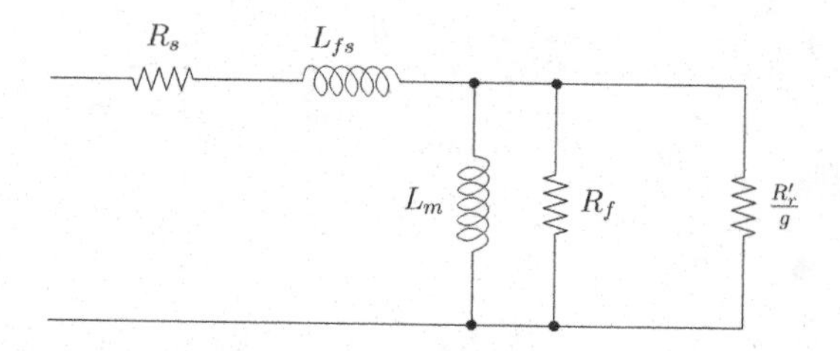


**Figure 2.15.** *Single-phase equivalent model of a squirrel-cage induction motor*

The diagram also shows the conversion of electromechanical energy, with the motional resistance $R'_r/g$, where $R'_r$ is the rotor resistance affecting the stator and $g$ the slide of the machine. We then see the stator resistance $R_s$ and the leakage inductance totaled in the stator $L_{fs}$, which shows an imperfect magnetic coupling between the stator coils and the squirrel cage located at the rotor. Finally, a resistance $R_f$ is classically added in parallel to the magnetizing inductance to take into account iron losses (by hysteresis and Foucault currents) in the ferromagnetic sheets in the machine.

In a way, an EMC is a simplified version of this representation, which eliminates aspects which are of secondary importance in HF: electromechanical energy conversion (the model no longer includes sliding). The motor is essentially seen as an iron–core coil (or, more accurately, three coupled coils). However, care must be taken to include non-negligible capacitive phenomena, with resonances and antiresonances which may, in practice, be observed across a frequency range of several tens of megahertz for a 1.5 kW motor (see [REV 03]). This thesis includes an basic model (see Figure 2.16) which takes into account these phenomena for frequencies of less than 1 MHz.

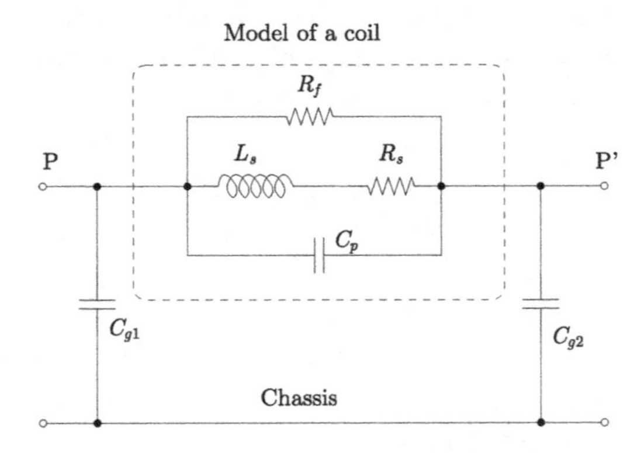


**Figure 2.16.** *Phase-by-phase model of a squirrel-cage inductance motor in relation to the chassis)*

This model includes the parasitic capacitances $C_{g1}$ and $C_{g2}$, which are the circulation pathways used by common-mode currents. A parasitic capacitance $C_p$ is located parallel to the classic model of an iron–core coil, made up of a series circuit ($L_s, R_s$) with the addition of a parallel resistance $R_f$ representing iron losses[10].

Quasi-separate identification of the three parasitic capacitances ($C_{g1}, C_{g2}$ and $C_p$) can be carried out using three distinct measurements as shown in Figure 2.17. Short circuits are successively established between:

– P and the chassis, for the first measurement, in order to identify $C_p + C_{g2}$;

– P' and the chassis, for the second measurement, in order to identify $C_{g1} + C_p$;

– P and P', for the final measurement, in order to identify $C_{g1} + C_{g2}$.

Beyond 1 MHz, this model ceases to be sufficient due to the appearance of multiple resonances and antiresonances. The model, as shown in Figure 2.16, is therefore replaced by a model using multiple parallel $R$, $L$ and $C$ cells, as shown in Figure 2.18. In this structure, the number of cells used $N_{cell}$ is equal to the number of observed resonances (note: $N_{cell}$ is equal to the number of resonances, not to the sum of the resonances and antiresonances).

Note that this model corresponds, in physical terms, to accounting for interturn capacitances, which cannot be globalized within a coil. From a mathematical perspective, the correlation between the model and experimental measurements obtained using impedance analysis can be

10 Strictly speaking, this resistance should be placed parallel to $L_s$ and not to the coil resistance $R_s$, which is purely representative of Ohmic voltage drops in the copper. As this is an equivalent model of a nonlinear device, the domain of validity is theoretically limited to a specific operating point; in practice, however, it is satisfactory for a wide range of frequencies.

established by noting that a parallel cell $(R_i, L_i, C_i)$ presents a resonance (high impedance) at the following frequency $f_{ri}$:

$$f_{ri} = \frac{1}{2\pi\sqrt{L_i C_i}} \qquad [2.5]$$

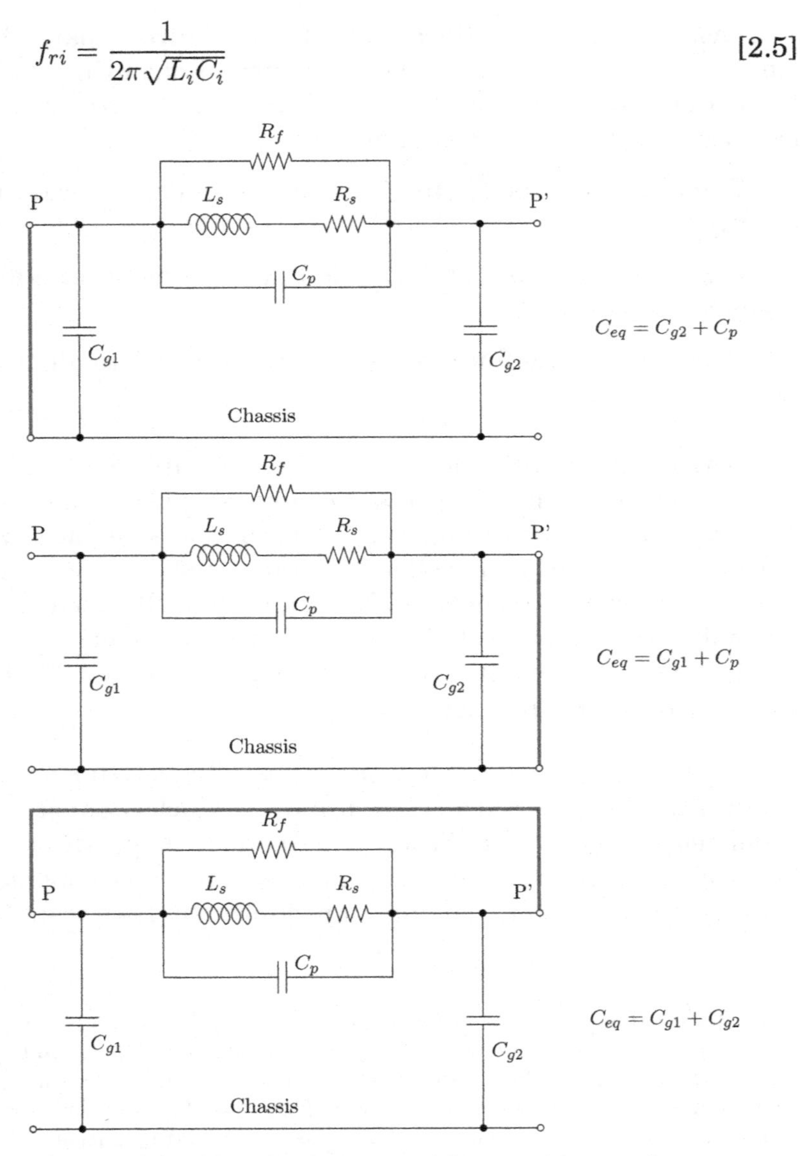


**Figure 2.17.** *Identification protocol for parasitic capacitances*

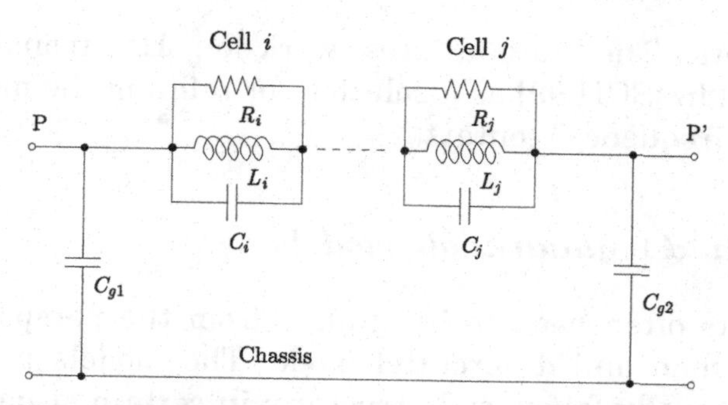


**Figure 2.18.** *Multicell model of an induction motor for high frequencies*

Antiresonances are obtained by intercellular association (i.e. between the inductance $L_i$ and the capacitance $C_j$). This gives an expression of the antiresonance frequency $f_{aij}$ of the form:

$$f_{aij} = \frac{1}{2\pi\sqrt{L_i C_j}} \quad [2.6]$$

The set of inductances introduced into each cell must verify the following equality:

$$\sum_{k=1}^{N_{cell}} L_i = L_s \quad [2.7]$$

where $L_s$ is the coil inductance observed at low frequencies. This inductance may then be split into multiple elements (in decreasing quantity order), and the capacitances of each cell are calculated in order to establish the resonance and antiresonance frequencies of the model, in accordance with the observations obtained through impedance analysis.

[REV 03] notes that this model does not accurately account for the "low"-frequency behavior of the coil, as the model presumes that inductances are constant, while phenomena such as the skin effect will modify these parameters. The author cites work on HF transformers presented in [SCH 99] as a solution for refining the model for this "low-frequency" context.

### 2.5.4. *T and* Π *quadrupole models*

Circuits often need to be studied from the perspective of both common and differential mode. The models presented above generally follow a Π structure; in certain situations, a T representation is preferable. A transformation should therefore be carried out; this is always possible in a linear context, although it is no longer possible to produce a circuit based on constant resistances, capacitances and inductances in this situation. In such cases, it is better to express the relationships between impedances, based on the diagram and notation shown in Figure 2.19. In these diagrams, the ground clearly corresponds to a motor chassis, the shielding of a cable or the heat sink of an electronic power converter (a variable speed drive, in the context of a power supply for an electrical machine).

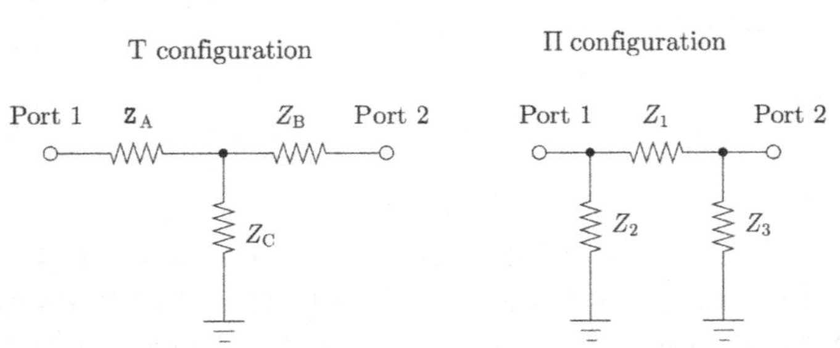


**Figure 2.19.** *T and* Π *models of quadrupoles*

REMARK 2.7.– Note that the T and Π quadrupoles proposed here for modeling purposes are dissymmetric. This is the most widespread configuration, although it is not always the case, and it is sometimes not possible to measure potential dissymmetry in practice.

The transformation between a T model and a Π model is known as Kennelly's theorem (this is the same transformation used to pass from star to delta coupling[11]). The transformation relationships are as follows:

$$\begin{cases} Z_A = \frac{Z_1 Z_2}{Z_1+Z_2+Z_3} \\ Z_B = \frac{Z_1 Z_3}{Z_1+Z_2+Z_3} \\ Z_C = \frac{Z_2 Z_3}{Z_1+Z_2+Z_3} \end{cases} \qquad [2.8]$$

And in the opposite direction:

$$\begin{cases} Z_1 = \frac{Z_A Z_B + Z_B Z_C + Z_C Z_A}{Z_C} \\ Z_2 = \frac{Z_A Z_B + Z_B Z_C + Z_C Z_A}{Z_B} \\ Z_3 = \frac{Z_A Z_B + Z_B Z_C + Z_C Z_A}{Z_A} \end{cases} \qquad [2.9]$$

In the case of symmetrical quadrupoles (i.e. for $Z_A = Z_B = Z_{AB}$ and $Z_2 = Z_3 = Z_{23}$), the following simplifications are used:

$$\begin{cases} Z_A = \frac{Z_1 Z_{23}}{Z_1+2Z_{23}} \\ Z_B = \frac{Z_1 Z_{23}}{Z_1+2Z_{23}} = Z_A \\ Z_C = \frac{Z_{23}^2}{Z_1+Z_2+Z_3} \end{cases} \qquad [2.10]$$

And in the opposite direction:

$$\begin{cases} Z_1 = \frac{Z_{AB}^2 + 2Z_{AB} Z_C}{Z_C} \\ Z_2 = \frac{Z_{AB}^2 + 2Z_{AB} Z_C}{Z_{AB}} \\ Z_3 = \frac{Z_{AB}^2 + 2Z_{AB} Z_C}{Z_{AB}} = Z_2 \end{cases} \qquad [2.11]$$

11 The T quadrupole case is clearly a specific form of the star configuration, while the Π case is a specific form of the star configuration.

The differential mode $Z_d$ and common-mode $Z_{mc}$ impedances that lead to perfectly decoupled behaviors can then be clearly identified:

$$Z_d = 2Z_A = \frac{2Z_1 Z_{23}}{Z_1 + 2Z_{23}} \quad [2.12]$$

and:

$$Z_{mc} = \frac{Z_2}{2} = \frac{Z_{AB}^2 + 2Z_{AB}Z_C}{2Z_{AB}} \quad [2.13]$$

Generally speaking, however, the common and differential modes cannot be separated. Mode coupling therefore occurs: a differential mode disturbance can lead to a common-mode disturbance and vice versa. For this reason, symmetrical structures are often preferred for filters, converters, cables and/or motors, although this is not an absolute rule.

## 2.6. Filtering

### 2.6.1. *Differential modes*

#### 2.6.1.1. *LC filtering*

Differential mode filtering is the most traditional filtering function, and has been encountered on several occasions in previous chapters:

– at the output of a rectifier bridge in Chapter 2 of Volume 2;

– at the output of switch-mode power supplies (such as the buck chopper or the forward power supply), seen in Chapters 1 and 2 of Volume 3.

In its simplest form, filtering in power electronics is carried out by an $LC$ circuit which constitutes a $2^{nd}$-order low-pass filter, which, in an ideal model, is non-dissipative (and in reality is only slightly dissipative). This is a key point in the creation of a static converter where efficiency needs to be maximized. However, this type of device needs to respond to certain constraints:

– avoid the generation of resonance or instability[12];

– filter interference effectively;

– not take up too much space;

– not cause excessive degradation of the overall efficiency of the converter.

We therefore have a constraint-based optimization problem, which will not be dealt with here; we will simply provide a reminder of the structure of the *LC* filter (Figure 2.20).

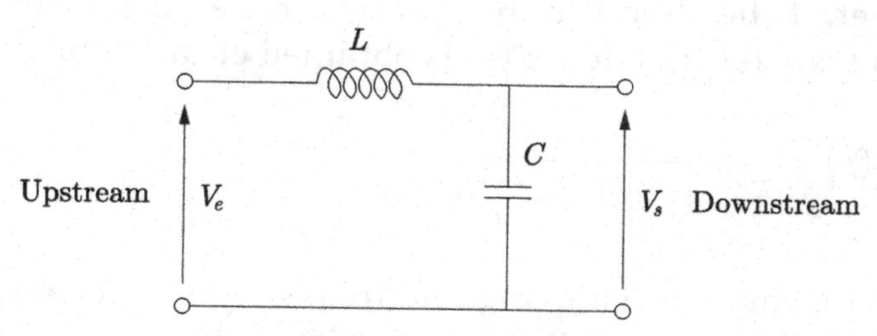


**Figure 2.20.** *Differential mode LC filter*

Note, however, that this structure is interesting from both upstream and downstream perspectives in terms of filtering:

– it smooths the current-experienced upstream, where the filter is perceived as a current source in the sense of power electronics;

– it smooths the voltage-experienced downstream, where the filter is perceived as a voltage source in the sense of power electronics.

12 This is unfortunately possible in a low dissipation circuit (with low damping).

The open transfer function $H_0(p)$ of this "$LC$" bridge may be established as:

$$H_0(p) = \frac{1}{1 + LCp^2} = \frac{1}{1 + \frac{p^2}{\omega_0^2}} \qquad [2.14]$$

This highlights the presence of a characteristic angular frequency (i.e. resonance frequency) $\omega_0 = \frac{1}{\sqrt{LC}}$ and a damping coefficient of zero (due to the absence of a $1^{st}$-order term in $p$ in the denominator). This filter is therefore at the stability borderline, with a resonance which is potentially infinite at angular frequency $\omega_0$. This resonance is limited in practice, on the one hand, by losses in the coil (iron and copper losses), but also in the capacitor (equivalent series resistance (ESR)). Moreover, if the circuit is charged by a resistance $R$ (parallel to $C$), a transfer function $H(p)$ is obtained of the form:

$$H(p) = \frac{1}{1 + 2z\frac{p}{\omega_0} + \left(\frac{p}{\omega_0}\right)^2} \qquad [2.15]$$

where the characteristic angular frequency is unchanged. A non-null damping coefficient $z$ also appears, which is expressed as:

$$z = \frac{R}{2}\sqrt{\frac{C}{L}} \qquad [2.16]$$

Another form of the transfer function is often encountered in publications on the subject, introducing the filter quality factor $Q$:

$$H(p) = \frac{1}{1 + Q\left(\frac{p}{\omega_0} + \frac{\omega_0}{p}\right)} \qquad [2.17]$$

where:

$$Q = \frac{1}{2z} = \frac{1}{R}\sqrt{\frac{L}{C}} \qquad [2.18]$$

#### 2.6.1.2. *The stability problem*

The stability problem associated with *LC* filters was mentioned in the previous section. At first glance, it seems surprising that a circuit of this type may be unstable. However, as demonstrated in calculating the transfer function $H_0$, the filter is at a stability limit when open. It stabilizes as soon as it is charged by a positive resistance. We may thus suppose that most of the time, a load-connected upstream of the filter will be dissipative; however, it is necessary to ensure that this will always be the case when an upstream converter is powering a machine. The problem is not only applicable to phases where energy is returned to the DC bus, if the filter is situated at the input terminals of a reversible converter, as seen in the first section; it may also occur in overspeed phases for machines with controlled excitation.

Classically, constant couple machines are controlled at low speeds, and constant power is then used at higher speeds. In the case of constant power operation, the characteristic of the "converter/machine" dipole in the plane $V(I)$ is a hyperbola. Linearization around an operating point shows a negative dynamic resistance, which is a potential source of instability and resonance in the DC bus.

### 2.6.2. *Common mode*

In the case of common-mode filtering, the element used needs to demonstrate low impedance in relation to the differential mode, and high impedance in relation to common mode. To do this, coupled inductances are used, with matching dots placed as shown in Figure 2.21.

In this configuration, in the presence of a differential current, the flux in the "outward" conductor and the flux in the "inward" conductor cancel out; for two currents circulating in the same direction (i.e. both entering through matching dots), the system will present high impedance. To establish an equation model of the voltage drop $\Delta V$ in the coils in common mode, we first write $I_A = I_B = I_{cm}/2$. Hence:

$$\Delta V = \frac{L+M}{2} \cdot \frac{dI_{cm}}{dt} \qquad [2.19]$$

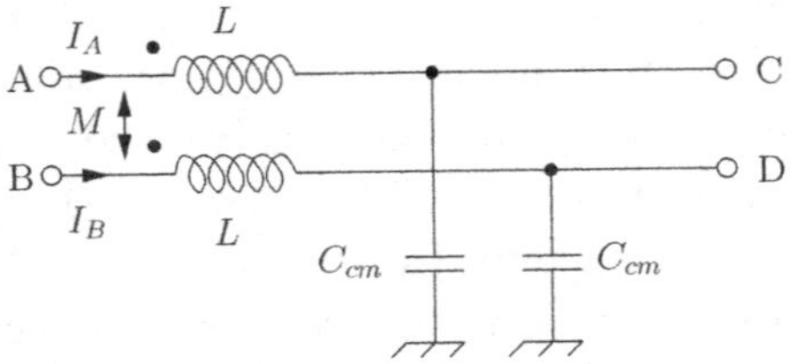

**Figure 2.21.** *Common-mode filter*

Note that the impedance experienced by this system, globally traversed by $I_A + I_B$, presents an equivalent inductance $\frac{L+M}{2}$ which is connected to a global capacitance $2C_{cm}$. This therefore constitutes a 2nd-order low-pass filter, for which the characteristic angular frequency can be calculated based on the results obtained for the differential filter $LC$. To be effective, a circuit of this type needs to use a magnetic circuit with high permeability (i.e. with no air gap), made from ferrite to allow it to operate at HFs. However, saturation induction is not a problem in this type of application, as common-mode currents are generally low (remembering that these are leakage currents).

A “differential mode filter + common-mode filter” system is not always found in wiring, as is often seen in the case of computer cables, such as USB cables. In this case (see Figure 2.22), a node is used to cover the cable, and therefore all of the wires, in order to produce a high mutual blocking common-mode interference.

### 2.6.3. *Limitations and design difficulties*

The filter diagrams shown in Figures 2.20 and 2.21 represent idealized structures. In Chapter 5 of Volume 1 [PAT 15a], we saw that real capacitors present an ESR or even a series inductance (ESL), which degrades the "short-circuit"-type behavior expected at high frequencies. Similarly, a real inductance presents a coil resistance (aggravated by the fact that the frequency increases with the skin effect), an iron-loss resistance (even if a ferrite, amorphous or nanocrystalline core is used: these are all suitable for use at HFs and with the components of EMC filters). Moreover, we have also seen in Figures 2.14 and 2.18, for the transformer and the inductance motor, respectively, that coils are characterized by parasitic capacitances between turns, but also (and especially) in relation to the magnetic circuit. These parasitic capacitances between coils and the magnetic core can be significant at filter operating frequencies when ferrites are used, as these materials are ceramic and present a high relative permittivity $\varepsilon_r$.

These limitations in real components reduce the efficiency of filters, and need to be taken into account during the design process [REV 03] in order to produce a satisfactory result and respect the disturbance frequency ranges imposed by current standards.

**Figure 2.22.** *Common-mode filter of a USB cable*

### 2.6.4. *Further reading*

The filters presented in this chapter (for both common and differential mode) are specific examples, suited to a situation where the impedance of the upstream dipole (source) is low, while the impedance of the downstream dipole (load) is high. Different filter variations are available for other contexts; a full list of commercially available filters may be found in [ESC 87].

## 2.7. Experimental aspects

### 2.7.1. *Line impedance stabilization networks (LISNs)*

An experimental phase is always required when analyzing conducted interference phenomena (in differential or common mode). In the case of applications connected to the EDF network (*Electricité de France*: the French mains network), it is hard to guarantee the reproducibility of tests and measurements due to fluctuations in network parameters, notably the impedance. While the frequency is precise (50 Hz $\pm$ 0.5 Hz), network equipment is designed to operate across root mean square (RMS) voltages varying from $0.85U_n$ to $1.1U_n$ ($U_n = 230$ V).

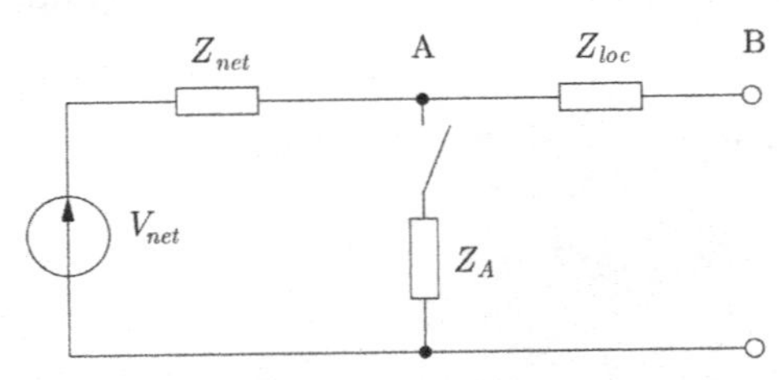


**Figure 2.23.** *Fluctuating impedance in the network*

This variability in voltage terms is linked to the consumption of loads connected to the network. Using a diagrammatic representation of the problem (see Figure 2.23), we see that when a load with impedance $Z_A$ is connected to point A, the Thévenin impedance obtained at point B will be different from that produced by the network alone. In practice, it is very difficult (if not impossible) to control the impedance of the network in an EMC test laboratory located in the vicinity of offices, with computers and printers which may be switched on, off, put on standby, etc., at any moment.

This difficulty can only be overcome by isolating the device under test (DUT) from the fluctuating network, and supplying a normalized impedance across a certain range of frequencies. This is used by installing an element known as an LISN between the network and the load. A diagram of an LISN is shown in Figure 2.24.

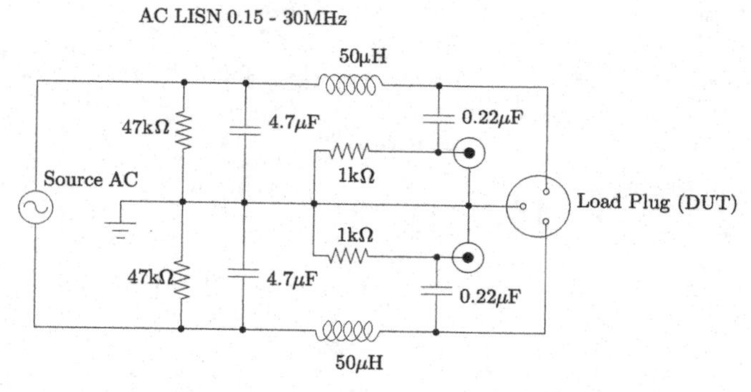


**Figure 2.24.** *LISN for an AC network*

REMARK 2.8.– The diagram shows an LISN for an AC network, but LISNs for DC networks are also available.

The two BNC measurement outputs shown in the diagram in Figure 2.24 and the photograph in Figure 2.25 (which includes four outputs) allow measurement of the potential differences between each of the two DUT power terminals and the ground (following high-pass filtering with a switching pulse of $\frac{1}{RC} = \frac{1}{0.22 \times 10^{-3}}$, giving a frequency of 724 Hz).

Note that these two signals do not directly supply information regarding the common and differential modes. To obtain this information, an active or passive device is used to calculate the difference and the sum of these two signals. Generally, the passive solution shown in the diagram of Figure 2.26 is preferred.

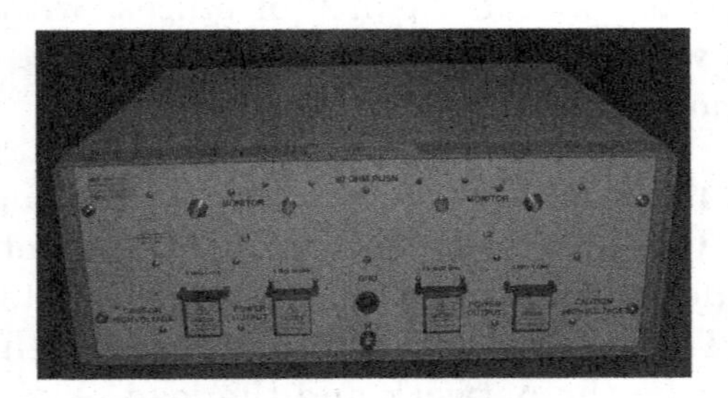

**Figure 2.25.** *Photograph of an AC LISN*

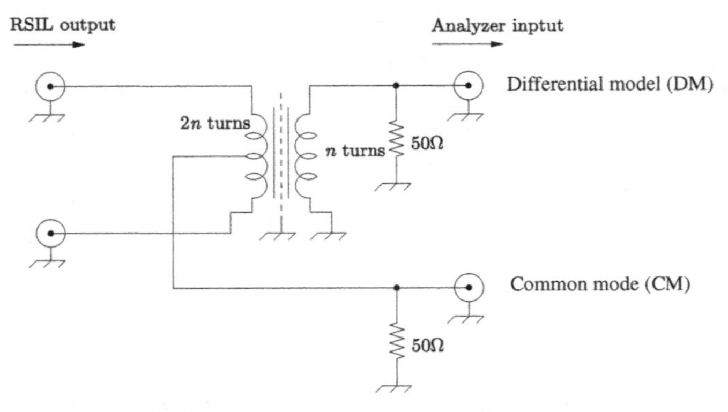


**Figure 2.26.** *Common- and differential-mode separator*

## 2.7.2. *Spectrum analyzer*

Spectrum analyzers (see Figure 2.27) are used alongside oscilloscopes for time/frequency analysis. The oscilloscope is essentially used to observe temporal variations of a signal, while the analyzer provides a frequency representation: it carries out a Fourier transform, which may be digital (using an FFT[13] algorithm, which may also be found in modern

13 Fast Fourier transform.

oscilloscopes), or analog (by frequency translation, via amplitude modulation, followed by peak, quasi-peak or RMS detection, based on the device and its configuration).

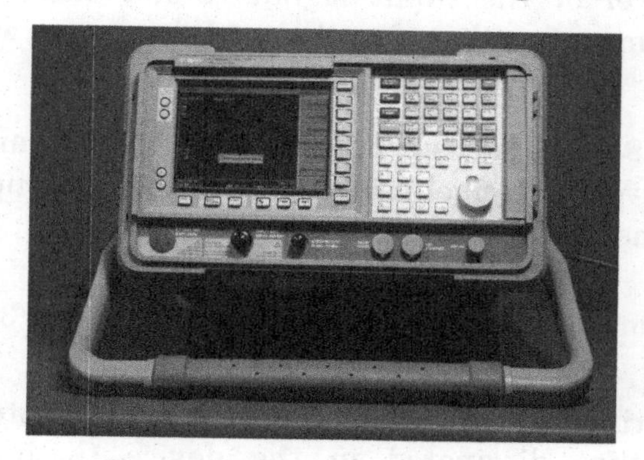

**Figure 2.27.** *Spectrum analyzer*

Thus, using a spectrum analyzer, we no longer consider the temporal evolution of a signal (which is no longer known), but simply the frequency content, which indicates whether the system evolves slowly (spectral content limited to low frequencies) or with rapid variation (shown by the presence of HF components). In the case of power electronics generally, and more particularly in EMC, special consideration is given to HF components, and the spectrum analyzer (and not the FFT function of an oscilloscope) constitutes an ideal tool for measuring HF interference.

However, there are certain restrictions concerning the use of this tool:

– the equipment is expensive and covers wide frequency ranges (up to 3 GHz for "basic" models);

– the bandwidth is limited at low frequencies, generally to 9 kHz, and additional payment is required for extension to lower frequencies;

– the device is often used in the RF domain, and is designed for 50 Ω sources;

– the level of the input signal cannot exceed a given threshold, meaning that a probe (active or otherwise) must be used.

REMARK 2.9.– The measurements $M$ produced by an analyzer are often given in decibel-milliwatt, and the reference value is therefore the milliwatt. Hence:

$$M\,[\mathrm{dBm}] = 10.\log\left(\frac{P\,[\mathrm{W}]}{10^{-3}}\right) = 10.\log\left(P\,[\mathrm{W}]\right) + 30 \qquad [2.20]$$

Considering the analyzer input voltage $U$, note that the power is that dissipated in the device (with an input impedance of 50 Ω). Hence:

$$M\,[\mathrm{dBm}] = 20.\log\left(U\,[\mathrm{V}]\right) + 13.01 \qquad [2.21]$$

Finally, this relationship may be inverted to obtain $U$ in volts from a measurement in dBm:

$$U\,[\mathrm{V}] = 10^{\frac{M[\mathrm{dBm}]-13.01}{20}} \qquad [2.22]$$

REMARK 2.10.– Measurements may also be given in dB$\mu$V. As the name indicates, this logarithmic unit uses the microvolt as a reference value:

$$M\,[\mathrm{dB}\mu\mathrm{V}] = 20.\log\left(\frac{U\,[\mathrm{V}]}{10^{-6}}\right) = 20.\log\left(U\,[\mathrm{V}]\right) + 120 \qquad [2.23]$$

The inversion of this relationship produces:

$$U\,[\mathrm{V}] = 10^{\frac{M[\mathrm{dB}\mu\mathrm{V}]-120}{20}} \qquad [2.24]$$

### 2.7.3. *Impedance analyzers*

An impedance analyzer (see Figure 2.28) may be used to characterize certain elements of an electronic system centered on a static converter.

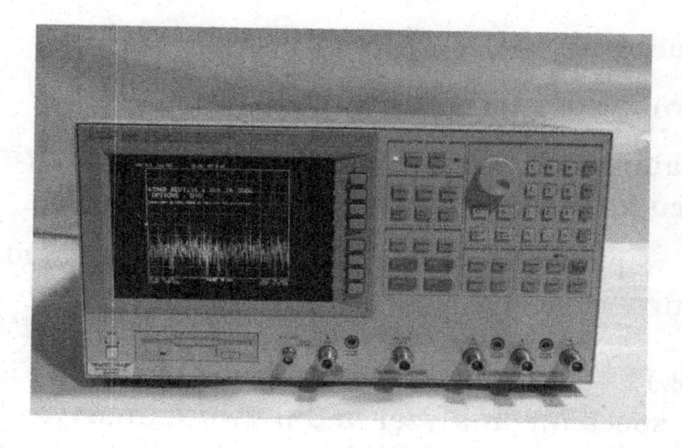

**Figure 2.28.** *Agilent 4396B impedance analyzer*

An impedance analyzer is perfectly suited to identifying the parameters (impedances) of a cable, transformer, electric motor or elementary components (capacitors, coils, etc). This tool allows precise measurement of the impedance of a linear dipole, in a more or less automatic fashion, across a wide range of frequencies. The complex result $\underline{Z}$ may be presented digitally in a variety of forms:

– Cartesian representation: real elements $R = \mathfrak{Re}[\underline{Z}]$ (resistance) and imaginary elements $X = \mathfrak{Im}[\underline{Z}]$ (reactance);

– polar representation: modulus $|\underline{Z}|$ and argument $\arg(\underline{Z})$.

Based on a given equivalent model ($R, C$ or $R, L$; series or parallel), the impedance analyzer is able to propose parameters. Finally, top-of-the-range impedance analyzers also automatically produce impedance graphs for a whole range of frequencies (with the potential to identify more complex moduli, typically of the $2^{nd}$-order, e.g. an $R, L, C$ series). Clearly, this type of postprocessing activity can be carried out in any case using calculation software by connecting the device to a computer (as in the case of all modern laboratory equipment).

Impedance meters are generally characterized by:

– their accuracy;

– their range of operating frequencies;

– their ability to operate with continuous polarization (e.g. useful when identifying batteries);

– their power level (measurement at high voltages/currents).

REMARK 2.11.– The impedance analyzer shown in Figure 2.28 also offers spectrum analyzer and network analyzer functions (the network analyzer will be discussed in the next chapter).

# 3

# Distributed Element Models

## 3.1. Aspects of electromagnetism

### 3.1.1. *Context and notation*

The aspects of electromagnetism presented below cover the case of a vacuum (with dielectric permittivity $\varepsilon_0 \simeq 8.85 \times 10^{-12}$F/m and magnetic permeability $\mu_0 = 4\pi \times 10^{-7}$T.m/A). Note that the following notation will be used:

– scalar quantities and operators will be noted in standard font (e.g. $\rho$, div, $V$, etc.);

– vector quantities and operators will be noted in bold (e.g. **j**, **E**, **D**, **H**, **B**, **curl**, etc.).

REMARK 3.1.– The differential operators used here are classic tools for electromagnetism, generally taught in the first years of university or in preparatory engineering courses. A variety of high-quality books are available for readers wishing to refresh their knowledge; [LUM 00] or [APP 02] are particularly recommended.

### 3.1.2. *Maxwell equations*

The basis for modern electromagnetism was established in 1864 by J.C. Maxwell, who established the four equations which bear his name. These equations are rooted in earlier work (by C.F. Gauss, A.M. Ampère, M. Faraday, etc). First, we note the Maxwell–Gauss equation linking the electrical field **E** to the electrical load density $\rho$:

$$\mathrm{div}\,\mathbf{E} = \frac{\rho}{\varepsilon_0} \quad [3.1]$$

Moreover, the vector field **D** may be introduced (the "electrical displacement" vector), defined as $\mathbf{D} = \varepsilon_0\mathbf{E}$; the previous equation can then be rewritten as:

$$\mathrm{div}\,\mathbf{D} = \rho \quad [3.2]$$

A second equation establishes the link between the magnetic field **H**, the load current density **j** and a displacement current density $\frac{\partial \mathbf{D}}{\partial t}$. This is the Maxwell–Ampère equation:

$$\mathbf{curl}\,\mathbf{H} = \mathbf{j} + \frac{\partial \mathbf{D}}{\partial t} \quad [3.3]$$

Note, as for the electric field, there is a connection between the magnetic field **H** and a second field, known as the magnetic induction (or magnetic flux density), **B**, which involves the magnetic permeability $\mu_0$:

$$\mathbf{B} = \mu_0\mathbf{H} \quad [3.4]$$

This equation may be used to rewrite [3.3] as follows:

$$\mathbf{curl}\,\mathbf{B} = \mu_0\mathbf{j} + \mu_0\varepsilon_0\frac{\partial \mathbf{E}}{\partial t} \quad [3.5]$$

The third Maxwell equation is generally known simply as the magnetic flux conservation equation:

$$\operatorname{div} \mathbf{B} = 0 \qquad [3.6]$$

Finally, the fourth equation (the Maxwell–Faraday equations) establishes the relationship between the electric field E and the temporal variations in the magnetic induction B:

$$\mathbf{curl}\, \mathbf{E} = -\frac{\partial \mathbf{B}}{\partial t} \qquad [3.7]$$

### 3.1.3. *Scalar and vector potentials*

The quantities handled here are space variables, which may be linked to the electrical quantities classically used in analyzing electrical circuits (voltage and current). The voltage $V$ between two conductors is linked to the electric field by a gradient:

$$\mathbf{E} = -\mathbf{grad}\, V \qquad [3.8]$$

which enables calculation of a potential difference between two points A and B using an integral along a path (between A and B) in E:

$$V(\mathrm{B}) - V(\mathrm{A}) = \int_{\mathrm{A}}^{\mathrm{B}} \mathbf{E} \cdot d\mathbf{l} \qquad [3.9]$$

Current $I$ may be obtained by calculating the flux of a current density j across a surface $\Sigma$:

$$I = \iint_{\Sigma} \mathbf{j} \cdot d\mathbf{s} \qquad [3.10]$$

The electric field (defined by a divergence, based on the Maxwell–Gauss equation) is therefore linked to a scalar potential at all spatial points. In the case of the magnetic field **B**, equation [3.6] is verified by the curl of this field, and the notion of vector potential **A** may be introduced, such that:

$$\mathbf{B} = \mathbf{curl}\,\mathbf{A} \quad [3.11]$$

Finally, note that the Maxwell–Gauss and magnetic flux equations are often known as initial condition equations in that, if these equations are satisfied at a given moment, they will always be satisfied as long as the two other equations are verified. These latter equations are dynamic equations due to the presence of a term in $\frac{\partial}{\partial t}$.

### 3.1.4. *Initial conditions*

The electrical charge conservation law (a basic principle behind electromagnetic theory) can be deduced from the Maxwell equations [3.2] and [3.3]. Let us begin by noting the local flux-divergence equation:

$$\operatorname{div}\mathbf{j} + \frac{\partial \rho}{dt} = 0 \quad [3.12]$$

Deriving equation [3.2] in relation to time gives:

$$\frac{\partial}{\partial t}\left[\operatorname{div}\mathbf{D}\right] = \operatorname{div}\frac{\partial \mathbf{D}}{\partial t} = \frac{\partial \rho}{\partial t} \quad [3.13]$$

Based on [3.3], it is known as:

$$\frac{\partial \mathbf{D}}{\partial t} = \mathbf{curl}\,\mathbf{H} - \mathbf{j} \quad [3.14]$$

$\frac{\partial \mathbf{D}}{\partial t}$ can therefore be replaced by this expression in [3.13]:

$$\operatorname{div}\left(\mathbf{curl}\,\mathbf{H} - \mathbf{j}\right) = \operatorname{div}\mathbf{curl}\,\mathbf{H} - \operatorname{div}\mathbf{j} = \frac{\partial \rho}{\partial t} \quad [3.15]$$

Moreover, the curl of a field is known to have a divergence of zero ($\text{div}\,\mathbf{curl} \equiv 0$). This results in the same charge conservation equation [3.12]. Consequently, equation [3.2] may be seen as an initial condition of equation [3.3], which conserves this property at all instants (QED).

Let us now consider equations [3.6] and [3.7]. This is done using a similar method to that used for the previous equations, deriving the "static" equation [3.6]:

$$\frac{\partial}{\partial t}\left[\text{div}\,\mathbf{B}\right] = \text{div}\,\frac{\partial \mathbf{B}}{\partial t} = 0 \qquad [3.16]$$

Equation [3.7] can then be used to establish an expression of $\frac{\partial \mathbf{B}}{\partial t}$:

$$\frac{\partial \mathbf{B}}{\partial \mathrm{t}} = -\mathbf{rot}\,\mathbf{E} \qquad [3.17]$$

Once again, equation [3.16] comes down to calculating the divergence of the curl of a field which is, once again, zero (QED).

## 3.2. Guided propagation

### 3.2.1. *Introduction*

Before considering the propagation of electromagnetic waves in free space, it is important to study waveguides, which are used to confine electromagnetic waves within a closed space. Waveguides can take a variety of forms:

– copper pipes (often with a silver coating) with either a round or rectangular cross-section;

– coaxial cables or two-wire lines (generally twisted);

– optical fibers.

In this case, we will consider "cable" type waveguides, also known as TEM waveguides, where the electric and magnetic fields are oriented perpendicular to the axis of the cable (and thus to the direction of propagation). In this context, the notions of voltage and current are used in relation to the wires. However, these quantities retain an irreducible spatio-temporal dimension, which is not classically seen in the localized constant circuits considered in the previous chapter. The voltage and current in the cable are denoted as $v(x,t)$ and $i(x,t)$, respectively (where $x$ is the point under consideration along the length of the cable).

### 3.2.2. *Coaxial cable parameters*

More detailed consideration will be given in this section to the simple case of a coaxial or shielded cable, the geometry of which is defined in Figure 3.1. This cable is made up of a conducting core of radius $r_1$ and a conducting shield with internal radius $r_2$ and external radius $r_3$.

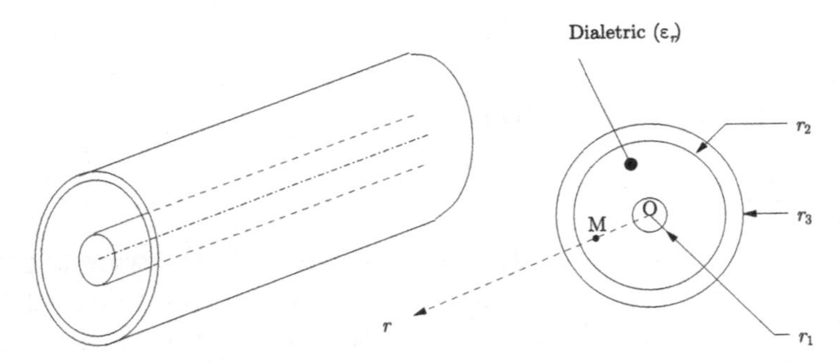


**Figure 3.1.** *Model of a coaxial cable*

As a starting point, the conductors will be presumed to be ideal (zero resistivity), while the space separating the core from the shield is considered to be filled with an insulating material of relative permittivity $\varepsilon_r$. Supposing that the cable is infinitely long, the cable may be studied as a cylindrical geometry problem (reduced to a 2D problem with a cutting plane following the direction of the cable) where the quantities only depend on the radius $r$ at the point under consideration.

Applying Gauss' theorem (i.e. the integral equation deduced from the Maxwell–Gauss equation[1]), it can be shown that:

$$E(r) = \frac{\rho_L}{2\pi r.\varepsilon_0\varepsilon_r} \; \forall r_1 < r < r_2 \qquad [3.18]$$

where $\rho_L$ is the linear density of charges in the core of the cable. This result may be supplemented by noting that the electric field is zero at all other points: not only in the conductors (presumed to be ideal), but also outside of the cable, where the conducting shield is charged with a linear density opposite to that of the core.

It is then easy to establish the expression of the potential difference $\Delta V$ between the two conductors to obtain an equation of form $\rho_L = C_L.\Delta V$, where $C_L$ is the linear capacitance of the cable. Thus:

$$C_L = \frac{2\pi\varepsilon_0\varepsilon_r}{\ln\left(\frac{r_2}{r_1}\right)} \qquad [3.19]$$

Ampère's theorem is then applied along a circular closed loop of radius $r$ centered on the axis of the cable in order to calculate the expression of the orthoradial induction field, with a modulus denoted as $B(r)$. To do this, a current $I$ is presumed to circulate within the central core, and return to the conducting shield in its entirety. Moreover, this current is presumed to be distributed in a uniform manner across the cut surface of the conductors. In these conditions, the magnetic field may be seen to be null outside of the cable (as

1 By applying the Green–Ostrogradski formula.

the sum of the outgoing and incoming currents is zero). However, there is a non-null magnetic field in both the dielectric and the conductors. Whatever the zone under consideration, the magnetic permeability is considered to have a value of $\mu_0$. Thus, the expression of $B(r)$ in the central conductor may be written as:

$$B\left(r\right)=\frac{\mu_0 I.r}{2\pi r_1^2}\ \forall r<r_1 \qquad [3.20]$$

Next, calculating the field in the dielectric, the cut current remains constant, while the length of the integration contour increases. This produces a field which decreases in terms of $1/r$:

$$B\left(r\right)=\frac{\mu_0 I}{2\pi r} \qquad [3.21]$$

Finally, when the integration contour enters the conducting shield, the magnetic field drops even more rapidly, reaching 0 when $r=r_3$:

$$B\left(r\right)=\frac{\mu_0 I}{2\pi r}\left(1-\frac{\left(r^2-r_2^2\right)}{\left(r_3^2-r_2^2\right)}\right) \qquad [3.22]$$

The energy $W_{\mathrm{Lmag}}$ stored per unit in the cable can then be integrated to give an equation of the form $W_{\mathrm{Lmag}}=\frac{1}{2}L_L.I^2$, where $L_L$ is the linear inductance of the cable. At high frequency (HF), note that the skin effect allows us to consider that the magnetic energy is essentially located in the dielectric. We thus obtain a simple expression of this linear inductance:

$$L_L=\frac{\mu_0}{2\pi}\ln\left(\frac{r_2}{r_1}\right) \qquad [3.23]$$

A line piece of length $dx$ can therefore be represented by an $LC$ circuit, where $L=L_L.dx$ and $C=C_L.dx$. This model corresponds to a lossless transmission line, LTL (leaving aside both the resistance in the conductors and losses in the dielectric). To produce a more accurate model, a resistance $R$ may simply be added in series with the inductance in order to take account of ohmic losses in the conductors, and a conductance $G$ is placed in parallel to the capacitor to model losses in the dielectric (see Figure 3.2). To complete the piece model, note that the resistance and the conductance may (as for $L$ and $C$) be expressed as a function of linear parameters, respectively denoted as $R_L$ and $G_L$, such that $R = R_L.dx$ and $G = G_L.dx$.

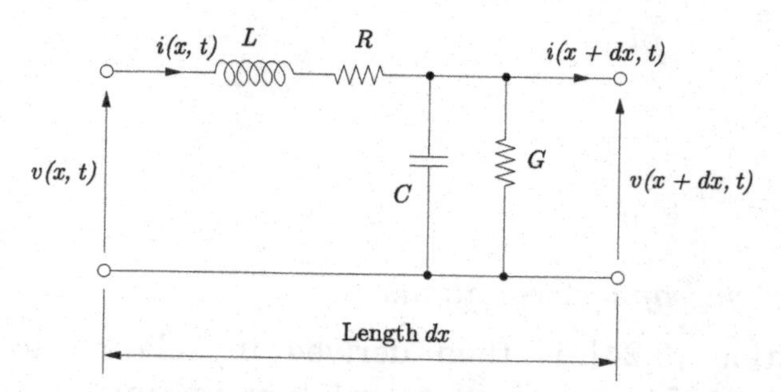


**Figure 3.2.** *RLCG model of a line piece*

### 3.2.3. *Line equations*

#### 3.2.3.1. *Equation model of a line slice*

The equation model of the line sliced is established by noting that the variation between the input and output voltage (i.e. the voltage variation per unit of length $\frac{\partial v}{\partial x}$) corresponds to the voltage drop at the terminals of the $R, L$ system (more accurately $R_L, L_L$ per unit of length):

$$\frac{\partial v}{\partial x} = -R_L i - L_L \frac{\partial i}{\partial t} \qquad [3.24]$$

In the same way, an equation is established to evaluate the difference between the input and output currents of the piece ($\frac{\partial i}{\partial x}$):

$$\frac{\partial i}{\partial x} = -G_L v - C_L \frac{\partial v}{\partial t} \quad [3.25]$$

This equation system is known as the line equations, and was established by O. Heaviside in the 1880s. In the case of a lossless line, the following simplified equations may be used:

$$\frac{\partial v}{\partial x} = -L_L \frac{\partial i}{\partial t} \quad [3.26]$$

and:

$$\frac{\partial i}{\partial x} = -C_L \frac{\partial v}{\partial t} \quad [3.27]$$

3.2.3.2. *Telegrapher's equations*

Equation [3.24] is then derived in relation to $x$ and equation [3.25] is derived in relation to time in order to obtain the Telegrapher's equation:

$$\frac{\partial^2 v}{\partial x^2} - L_L C_L \frac{\partial^2 v}{\partial t^2} - L_L G_L \frac{\partial v}{\partial t} - -R_L G_L \frac{\partial v}{\partial t} - R_L G_L v = 0 \quad [3.28]$$

The same equation is obtained for the current (simply replacing $v$ with $i$).

These equations are simplified in the case of a lossless line:

$$\frac{\partial^2 v}{\partial x^2} - L_L C_L \frac{\partial^2 v}{\partial t^2} = 0 \quad [3.29]$$

and:

$$\frac{\partial^2 i}{\partial x^2} - L_L C_L \frac{\partial^2 i}{\partial t^2} = 0 \quad [3.30]$$

3.2.3.3. *Harmonic state*

In the case of harmonic state (or sinusoidal steady-state), complex solutions $\underline{v}(x,t)$ and $\underline{i}(x,t)$ with separable variables should be sought, in order to write:

$$\underline{v}(x,t) = \underline{V}(x).e^{j\omega t} \quad [3.31]$$

and:

$$\underline{i}(x,t) = \underline{I}(x).e^{j\omega t} \quad [3.32]$$

Complex amplitudes $\underline{V}(x)$ and $\underline{I}(x)$ therefore need to be defined in order to solve the problem. The propagation equations lead to the establishment of equations of the form:

$$\underline{V}(x) = A.e^{\gamma x} + B.e^{-\gamma x} \quad [3.33]$$

and:

$$\underline{I}(x) = \frac{1}{Z_c}\left(A.e^{\gamma x} - B.e^{-\gamma x}\right) \quad [3.34]$$

where parameters $\gamma$ and $Z_c$ (respectively, the propagation constant and the characteristic impedance of the cable in $\Omega$) are expressed as:

$$\gamma = \left[(R_L + jL_L\omega).(G_L + jC_L\omega)\right]^{1/2} = \alpha + j\beta \quad [3.35]$$

and:

$$Z_c = \left(\frac{R_L + jL_L\omega}{G_L + jC_L\omega}\right)^{1/2} \quad [3.36]$$

In a lossless context, simplified expressions of these two parameters are used (denoted as $\gamma^{\mathrm{LTL}}$ and $Z_c^{\mathrm{LTL}}$):

$$\gamma^{\mathrm{LTL}} = \left[jL_L\omega.jC_L\omega\right]^{1/2} = j\underbrace{\omega\sqrt{L_L C_L}}_{\beta^{\mathrm{LTL}}} \quad [3.37]$$

and:

$$Z_c^{\mathrm{LTL}} = \sqrt{\frac{L_L}{C_L}} \qquad [3.38]$$

## 3.2.4. *Impedance of a line piece*

### 3.2.4.1. *Impedance reference planes*

An impedance $\underline{Z}(x)$ should now be defined for a given point $x$ in the line (i.e. a reference plane[2]) which is simply the ratio between $\underline{V}(x)$ and $\underline{I}(x)$. This gives the following relationship:

$$\underline{Z}(x) = \frac{\underline{V}(x)}{\underline{I}(x)} = Z_c \cdot \frac{A.\mathrm{e}^{\gamma x} + B.\mathrm{e}^{-\gamma x}}{A.\mathrm{e}^{\gamma x} - B.\mathrm{e}^{-\gamma x}} \qquad [3.39]$$

Full solution of the propagation problem in the cable implies determining coefficients $A$ and $B$. We will now consider this point in greater detail, noting that, as for any partially derived equation, the telegrapher's equation requires the use of boundary conditions in order to be solved. In this case, these conditions are obtained by characterizing the elements placed at the extremities of the cable.

REMARK 3.2.– Position $x$ presumes that a reference point has been established; this may be one of the two extremities. For the remainder of this chapter, we will consider that position $x = 0$ corresponds to the load, which is therefore considered as the reference point. Note, however, that this choice is completely arbitrary[3].

### 3.2.4.2. *Integration of the line between a source and a load*

Let the load impedance be denoted as $Z_L$, as demonstrated in the previous section; this load is taken as a reference point,

2 An important notion when calibrating a network analyzer, as presented in section 3.2.5.3.

3 However, this choice is most widely used in publications on the subject.

and so we will consider that it is situated at $x = 0$. Therefore, $x$ increases toward the source.

The source is modeled as an equivalent Thévenin circuit, i.e. an ideal voltage source $\underline{e}(t) = E.e^{j\omega t}$ placed in series with an impedance $Z_S$. The diagram corresponding to this study is shown in Figure 3.3.

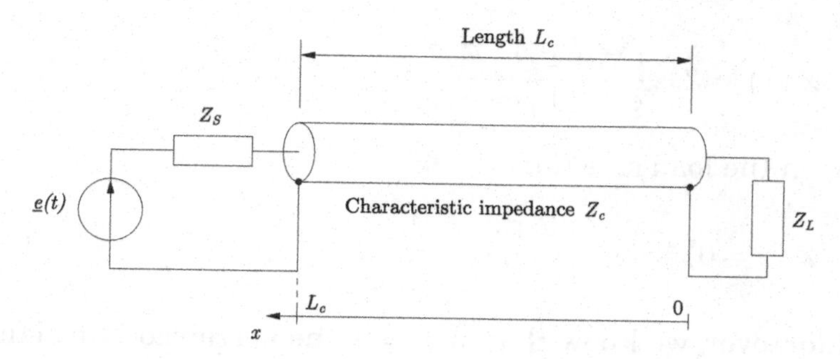


**Figure 3.3.** *Complete line (source + cable + load)*

On the load side, $Z(x = 0)$ may be calculated based on equation [3.39]:

$$Z\left(x=0\right) = Z_c \cdot \frac{A+B}{A-B} \qquad [3.40]$$

Considering the form of solutions $\underline{V}(x)$ and $\underline{I}(x)$, coefficients $A$ and $B$ can be seen to correspond to the amplitude of the voltages (and currents) circulating in the opposite direction: in the direction of decreasing $x$ (according to our convention), propagation occurs from the source toward the load, something known as an incident wave. In the direction of increasing $x$, movement occurs from the load toward the source: in these conditions, as the load is passive, the wave may be seen to be reflected by the load.

A reflection coefficient $\Gamma_L$ may then be easily introduced (at load level), defined as:

$$\Gamma_L = \frac{B}{A} \qquad [3.41]$$

The expression of $Z(x)$ may then be rewritten to include this new parameter:

$$\underline{Z}(x) = Z_c \cdot \frac{e^{\gamma x} + \Gamma_L e^{-\gamma x}}{e^{\gamma x} - \Gamma_L e^{-\gamma x}} \qquad [3.42]$$

and, on the load side (for $x = 0$):

$$Z(x = 0) = Z_c \cdot \frac{1 + \Gamma_L}{1 - \Gamma_L} \qquad [3.43]$$

Moreover, we know that at $x = 0$, the perceived impedance must be that of the load ($Z_L$). Hence:

$$Z_c \cdot \frac{1 + \Gamma_L}{1 - \Gamma_L} = Z_L \qquad [3.44]$$

As impedances $Z_c$ and $Z_L$ are known, this reflection coefficient $\Gamma_L$ can be expressed on the load side in the following way:

$$\Gamma_L = \frac{Z_L - Z_c}{Z_L + Z_c} \qquad [3.45]$$

REMARK 3.3.– Generally speaking, a reflection coefficient $\Gamma(x)$ may be defined at any given point in the line. Coefficient $\Gamma_L$ is therefore a specific instance for $x = 0$, and the general expression is as follows:

$$\Gamma(x) = \Gamma_L . e^{-2\gamma x} \qquad [3.46]$$

This result can then be used to calculate the impedance due to a "load + cable piece" system. Equation [3.42] can then be reformulated as:

$$\underline{Z}(x) = Z_c \cdot \frac{Z_L + Z_c \tanh(\gamma x)}{Z_c - Z_L \tanh(\gamma x)} \qquad [3.47]$$

where $\tanh(\cdot)$ is the hyperbolic tangent function, defined as:

$$\tanh(x) \triangleq \frac{\sinh(x)}{\cosh(x)} = \frac{e^x - e^{-x}}{e^x + e^{-x}} \qquad [3.48]$$

In the case of a lossless transmission line, equation [3.47] becomes:

$$\underline{Z}^{\mathrm{LTL}}(x) = Z_c \cdot \frac{Z_L + jZ_c \tan\left(\beta^{\mathrm{LTL}} x\right)}{Z_c - jZ_L \tan\left(\beta^{\mathrm{LTL}} x\right)} \qquad [3.49]$$

This result shows that the impedance is a complex quantity which is subject to periodic modifications, as the "tangent" function is $\pi$-periodic. The spatial periodicity $\Delta x$ of $\underline{Z}^{\mathrm{LTL}}(x)$ can therefore be obtained using the equation:

$$\beta^{\mathrm{LTL}} \Delta x = \pi \qquad [3.50]$$

hence:

$$\Delta x = \frac{\pi}{\beta^{\mathrm{LTL}}} = \frac{\pi}{\omega\sqrt{L_L C_L}} = \frac{c'}{2f} \qquad [3.51]$$

where $c'$ is defined as the wave propagation speed in the cable. Ratio $c'/f$ is the spatial period (or wavelength $\lambda$) of the wave with frequency $f$ propagating at speed $c'$. The spatial periodicity of the impedance is thus $\lambda/2$.

#### 3.2.4.3. *Impedance matching*

In the previous section, the impedance of a "cable piece + load" system was seen to present a periodic evolution based on the length $x$ of the piece with a period $\lambda/2$. However, note that this behavior can tend toward a trivial case, where $\underline{Z}(x)$ becomes constant for all values of $x$. Taking equation [3.42] and using $\Gamma_L = 0$, we obtain:

$$\underline{Z}(x) = Z_c \, \forall x \tag{3.52}$$

Condition $\Gamma_L = 0$ corresponds to the equation:

$$\frac{Z_L - Z_c}{Z_L + Z_c} = 0 \tag{3.53}$$

and thus to the case where the load impedance $Z_L$ is equal to the characteristic impedance $Z_c$ of the cable. In this case, there is impedance matching between the load and the cable. This corresponds to nonreflection of the wave transmitted by the cable on the load side: all of the power supplied by the source is transferred to the load.

Impedance matching on the source side may be considered in the same way. To do this, note that the impedance perceived by the source is $\underline{Z}(x = L_c)$. In the general case, this impedance is expressed as:

$$\underline{Z}(L_c) = Z_c \cdot \frac{Z_L + Z_c \tanh(\gamma L_c)}{Z_c - Z_L \tanh(\gamma L_c)} \tag{3.54}$$

For an LTL, this expression becomes:

$$\underline{Z}^{\mathrm{LTL}}(L_c) = Z_c \cdot \frac{Z_L + jZ_c \tan(\gamma^{\mathrm{LTL}} L_c)}{Z_c - jZ_L \tan(\gamma^{\mathrm{LTL}} L_c)} \tag{3.55}$$

A new reflection coefficient $\Gamma_S$ can then be defined on the source side, replacing $Z_c$ by $Z_S$ and $Z_L$ by $\underline{Z}(L_c)$ in the expression of $\Gamma_L$, giving:

$$\Gamma_S = \frac{\underline{Z}(L_c) - Z_S}{\underline{Z}(L_c) + Z_S} \quad [3.56]$$

As for the load, impedance matching occurs if $\Gamma_S = 0$, and thus if $\underline{Z}(L_c) = Z_S$.

In conclusion, a fully matched system is obtained if the impedance of the source, the cable (characteristic impedance) and the load are identical. Clearly, this type of system may also be partially matched, or totally unmatched.

3.2.4.4. *Power and impedance matching*

The concept of impedance matching is crucial when considering the power extracted from an impedant source connected to a charge. The notion of impedance matching is therefore not specifically linked to wave propagation in a cable; it also applies to a source – load system, as shown in Figure 3.3, where the two elements are connected directly (i.e. by a cable of negligible length $L_c$). In these conditions, a voltage divider bridge is obtained. Noting the voltage at the load terminals as $\underline{v}_L$, we obtain:

$$\underline{v}_L = \frac{Z_L}{Z_S + Z_L} \cdot \underline{e} \quad [3.57]$$

The current circulating in the "source $\underline{e}$ + $Z_S$ + $Z_L$" loop can also be expressed as:

$$\underline{i} = \frac{\underline{e}}{Z_S + Z_L} \quad [3.58]$$

The power $\mathcal{P}$ in a dipole (in this case, the load) as a function of the associated complex voltages and currents is written as:

$$\mathcal{P} = \Re\mathrm{e}\left[\underline{v}_L.\underline{i}\right] = \frac{1}{2}\left(\underline{v}_L.\underline{i}^* + \underline{v}_L^*.\underline{i}\right) \qquad [3.59]$$

Giving the following result:

$$\mathcal{P} = \frac{\Re\mathrm{e}\left[Z_L\right].\left|\underline{e}\right|^2}{\left|Z_S + Z_L\right|^2} \qquad [3.60]$$

This result may be used by introducing the real and imaginary parts of impedances $Z_S$ and $Z_L$:

$$\begin{cases} Z_S = R_S + jX_S \\ Z_L = R_L + jX_L \end{cases} \qquad [3.61]$$

This gives a new expression of the power:

$$\mathcal{P} = \frac{R_L.\left|\underline{e}\right|^2}{\left(R_S + R_L\right)^2 + \left(X_S + X_L\right)^2} \qquad [3.62]$$

Without needing to calculate the extremum in the space $(R_L, X_L)$, it is easy to see that the second term of the denominator can be canceled out for $X_L = -X_S$. This is the minimum value of this term; $X_S$ and $X_L$ are only involved in the expression of $\mathcal{P}$. The problem then consists of calculating the extremum of the reduced expression:

$$\mathcal{P} = \frac{R_L.\left|\underline{e}\right|^2}{\left(R_S + R_L\right)^2} \qquad [3.63]$$

This expression of $\mathcal{P}$ can then be derived in relation to $R_L$ to obtain the desired result:

$$\frac{\partial \mathcal{P}}{\partial R_L} = \frac{\left|\underline{e}\right|^2\left(R_S + R_L\right)^2 - 2R_L\left|\underline{e}\right|^2\left(R_S + R_L\right)}{\left(R_S + R_L\right)^4} \qquad [3.64]$$

This expression is canceled by canceling the numerator. This gives the following expression:

$$R_S^2 - R_L^2 = (R_S + R_L)(R_S - R_L) = 0 \quad [3.65]$$

in which impedance matching occurs when $R_L = R_S$.

The result obtained earlier therefore needs to be corrected as, generally speaking, impedance matching is not truly obtained for identical impedances (except in the case of an LTL for which the characteristic impedance is purely real), but, in fact, for:

$$Z_L = Z_S^* \quad [3.66]$$

3.2.4.5. *Standing-waves*

The spatial periodicity of the impedance $\underline{Z}(x)$ has already been discussed. This periodicity has an effect on the amplitude of the voltage and of the current ($\underline{V}(x)$ and $\underline{I}(x)$ respectively). In cases of unmatched impedance, the amplitude of sinusoidal signals observed in the line is dependent on the position $x$ under consideration: as in the case of vibrating strings, the amplitude reaches maximum and minimum values. This phenomenon is known as a standing-wave.

For illustrative purposes, let us consider the simple case of an LTL, with a matched source ($Z_S = Z_c$) at one extremity and a short circuit ($Z_L = 0$) at the other extremity. The value of the reflection coefficient on the load side is easy to calculate:

$$\Gamma_L = -1 \quad [3.67]$$

When calculating the impedance due to the short circuit in a line piece of length $\lambda/2$, the impedance is already known to be zero. However, when calculating this impedance for $x = \lambda/4$, a theoretical value of infinity is obtained:

$$\underline{Z}\,(x = \lambda/4) = \underline{Z}^{\mathrm{LTL}}\,(L_c) = jZ_c \tan\left(\frac{\beta^{\mathrm{LTL}}\lambda}{4}\right)$$
$$= jZ_c \tan\left(\frac{\pi}{2}\right) \to \infty \qquad [3.68]$$

The "piece + load" system may therefore be replaced by an open circuit. However, the source matched to the rest of the cable may be seen as the source alone. Applying the vector divider bridge formula, the voltage at the point under consideration is seen to be equal to the open e.m.f. of the source.

This phenomenon is repeated all along the cable. Measurements show that voltage maxima occur every $\lambda/2$; between these maxima, the voltage is canceled at points where the impedance cancels out. This is identical to the classic behavior of a vibrating string, with nodes and antinodes in the amplitude of oscillation.

REMARK 3.4.– This phenomenon also occurs in the case of infinite load impedance and, more generally, in all unmatched cases. However, when the load is matched, the amplitude of oscillation is constant along the whole length of the cable (and equal to the amplitude of the voltage $\underline{e}$ divided by 2).

The standing-wave ratio (SWR) is a parameter widely used to characterize the matched or mismatched nature of a "cable + load" system. This is simply the ratio between the maximum and minimum observable voltage amplitudes in the line:

$$\mathrm{SWR} = \frac{|A| + |B|}{|A| - |B|} = \frac{1 + |\Gamma\,(x)|}{1 - |\Gamma\,(x)|} = \frac{1 + \Gamma_L}{1 - \Gamma_L} \qquad [3.69]$$

In the case of an LTL:

$$\mathrm{SWR} = \frac{Z_L}{Z_c} \qquad [3.70]$$

This coefficient is equal to 1 in the case of a matched system. In the case of a short circuit, the coefficient is equal to zero; it tends toward infinity when the cable is in an open circuit at its extremity.

3.2.4.6. *The Smith chart*

The formulas presented in the previous sections make it possible to carry out all of the calculations required to understand the behavior of a transmission line, calculate impedances, reflections, etc.

However, a graphical tool may be used to replace all of these formulas by geometric operations (specifically, rotations). This tool is known as the Smith chart, and is shown in Figure 3.4.

Using this tool, a complex impedance may be represented using a network of "iso-resistance" and "iso-reactance" circles, then subjected to a transformation through a cable piece by rotation around the center of the main circle in the chart. Figure 3.4 shows indications showing the required sense of rotation, depending on whether movement is made toward the generator (clockwise) or toward the load (counter-clockwise). Note, moreover, that the polar gradations on the edge of the chart are given as fractions of the wavelength, and a full turn has a value of 0.5$\lambda$. This is perfectly coherent with the results established in equation [3.51], which indicated that the impedance of the load due to a line piece of length $x$ evolves with a spatial periodicity of $\lambda/2$.

The chart corresponds to a complex plane, with an origin at the center of the principle circle; it is not an impedance graph, but rather a graph showing the (complex) reflection coefficient $\Gamma$. The origin of the reference frame corresponds to zero reflection, which is generally associated with an impedance of $50\,\Omega$ (the characteristic impedance generally encountered in radio frequency (RF) with coaxial cables and measurement instruments: generators, oscilloscopes, spectrum analyzers, network analyzers, etc.). This complex plane is supplemented by a chart, made up of a network of curves establishing specific values for the real and imaginary parts of the normalized impedance $z = Z(x)/Z_c$, or, more accurately, according to $2x/\lambda$ (the normalized distance based on a half-wavelength).

3.2.4.7. *Impedance matching in power electronics*

Impedance matching is not a major aim in power electronics, and, in this context, is unnatural. While impedance matching may seem interesting in terms of power transfer, based on our previous discussion, we should remember that this study was based on an imposed impedant source with the aim of extracting the highest possible power level by matching the load. Note, however, that in this configuration, the extracted power is maximized, but at the expense of significant losses in the source: the power dissipated in the source in this case is equal to the power transferred to the load. Clearly, in the context of an antenna receiving a cellular telephone signal, this is not problematic, and is, moreover, essential for satisfactory signal retrieval; however, in power electronics, this situation (with an efficiency level of 0.5) is unacceptable, as loss reduction is key.

Low-impedance converters have been developed which can only be used at operating points far from a matched situation; these components would be unable to withstand matching. For illustrative purposes, note that switch-mode power supplies for MOSFET switches are widely used with a resistance of around $10\,\text{m}\Omega$ (or less) at a few volts (e.g. $5\,\text{V}$). These switches may be easily used for switched currents of a few tens of amps (on condition that they are cooled correctly);

however, a rapid calculation shows that in the case of matching, the current circulating in the switch would no longer be in the tens of amps ($5\,\text{V}/0.02\,\Omega = 250\,\text{A}$).

Finally, note that reflection and standing-wave phenomena are to be expected in power electronic devices in some corresponding geometrical and time/frequency scales (for example when the dimensions of the cables are significant in relation to the wavelength).

REMARK 3.5.– As power electronic devices present high levels of mismatching, it may be difficult to use the Smith chart to replace calculations (or, if the chart is used, the results will not be particularly accurate as the networks of "iso-resistance" and "iso-reactance" curves are limited to an insufficiently wide range of values). However, this simple tool enables a clearer understanding of these phenomena, and is particularly useful for teaching purposes.

### 3.2.5. *Quadripoles and "S" parameters*

#### 3.2.5.1. *Definitions*

The notion of the quadripole was introduced in the previous chapter in the context of a "lumped parameters circuit" approach, but is still relevant in the case of "distributed elements model", such as transmission lines. In this context, the impedance, admission and transfer matrices are replaced by the "S" parameter matrix (for Scattering parameters) linking the incident and reflected waves constituting input and output for the cable (or, more generally, any quadripole).

These waves are normalized in relation to the characteristic impedance of the cables. More precisely, if we consider the quadripole constituted by the line piece between $x_1$ and $x_2$, voltages $\underline{V}(x_1)$ and $\underline{V}(x_2)$ are defined as follows:

$$\begin{cases} \underline{V}(x_1) = A.\mathrm{e}^{\gamma x_1} + B.\mathrm{e}^{-\gamma x_1} \\ \underline{V}(x_2) = A.\mathrm{e}^{\gamma x_2} + B.\mathrm{e}^{-\gamma x_2} \end{cases} \qquad [3.71]$$

**The Complete Smith Chart**

Black Magic Design

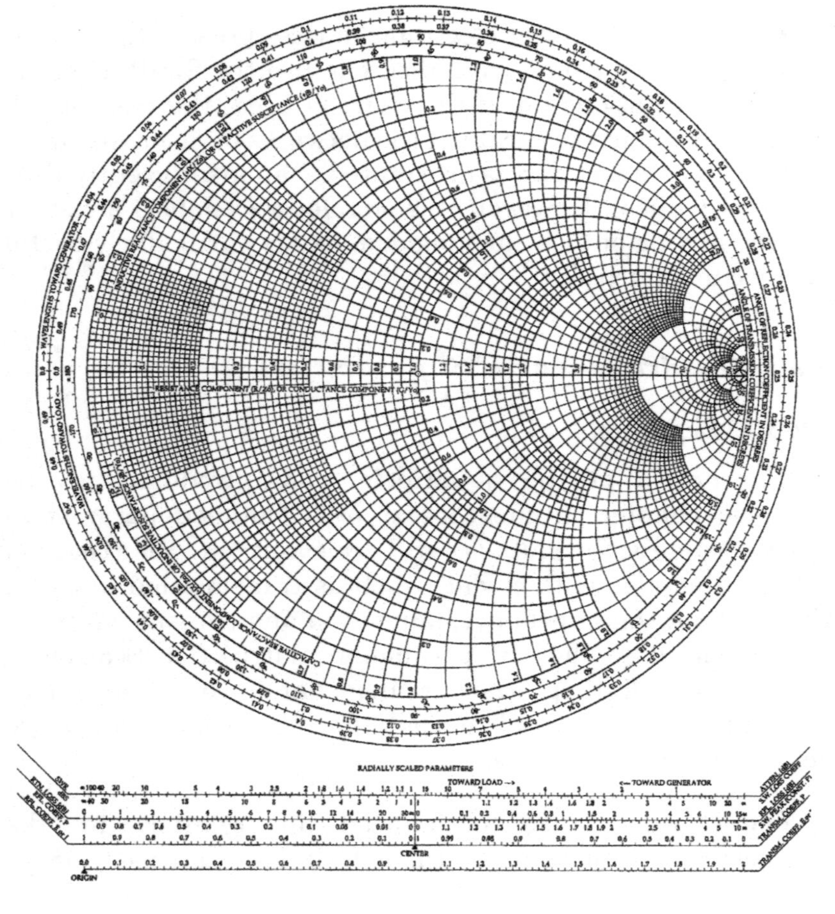


**Figure 3.4.** *The Smith chart*

Using the initial convention ($x = 0$ at the load and $x > 0$ toward the source), $x_1$ is therefore considered to be greater than $x_2$, and we note that terms in $A.\mathrm{e}^{\gamma x}$ evolve from the source toward the load, while the terms in $B.\mathrm{e}^{-\gamma x}$ evolve in the opposite direction. In these conditions, the incident and reflected waves at the two ports of the line-piece quadripole may be clearly identified:

– $A.\mathrm{e}^{\gamma x_1}$ corresponds to an incident wave at port 1 (localized in $x_1$);

– $B.\mathrm{e}^{-\gamma x_1}$ corresponds to a reflected wave at port 1;

– $A.\mathrm{e}^{\gamma x_2}$ corresponds to a reflected wave at port 2 (localized at $x_2$) ;

– $B.\mathrm{e}^{-\gamma x_2}$ corresponds to an incident wave at port 2.

The normalized waves linked to the "S" parameters are simply the components which have just been defined, up to a coefficient $\frac{1}{\sqrt{Z_c}}$:

– incident wave $a_1$ at port 1:

$$a_1 = \frac{A.\mathrm{e}^{\gamma x_1}}{\sqrt{Z_c}} \tag{3.72}$$

– reflected wave $b_1$ at port 1:

$$b_1 = \frac{B.\mathrm{e}^{-\gamma x_1}}{\sqrt{Z_c}} \tag{3.73}$$

– incident wave $a_2$ at port 2:

$$a_2 = \frac{B.\mathrm{e}^{-\gamma x_2}}{\sqrt{Z_c}} \tag{3.74}$$

– reflected wave $b_2$ at port 2:

$$b_2 = \frac{A.\mathrm{e}^{\gamma x_2}}{\sqrt{Z_c}} \tag{3.75}$$

3.2.5.2. *Uses of "S" parameters*

Parameters $S_{ij}$ are quadripole parameters which can be measured using a network analyzer (see section 3.2.5.3). Their utility may be called into question, as the notions of impedance and admittance also exist in this HF context. In fact, the equations of a quadripole with the coefficients of matrix $(Z)$ and matrix $(Y)$ may simply be used to highlight an experimental identification approach:

$$\begin{cases} V_1 = Z_{11}.I_1 + Z_{12}.I_2 \\ V_2 = Z_{21}.I_1 + Z_{22}.I_2 \end{cases} \qquad [3.76]$$

and:

$$\begin{cases} I_1 = Y_{11}.V_1 + Y_{12}.V_2 \\ I_2 = Y_{21}.V_1 + Y_{22}.V_2 \end{cases} \qquad [3.77]$$

The approach used to identify an impedance (for example $Z_{11}$) therefore consists of opening a circuit ($I_2 = 0$ to identify $Z_{11}$):

$$Z_{11} = \left.\frac{V_1}{I_1}\right|_{I_2=0} \qquad [3.78]$$

In the case of an admittance matrix (for example $Y_{11}$), a port must be short-circuited ($V_2 = 0$, to identify $Y_{11}$):

$$Y_{11} = \left.\frac{I_1}{V_1}\right|_{V_2=0} \qquad [3.79]$$

In HF, short circuits are particularly difficult to establish (due to the presence of parasitic inductances), and it is even harder to maintain a purely open circuit (due to parasitic capacitances). In these conditions, while matrices $(Z)$ and $(Y)$ are theoretically still valid and usable, they cannot be identified experimentally. "S" parameters are therefore preferred, as a reflected wave will be canceled in the case of connection of a matched load (which is much easier to obtain accurately in practice than a short-circuit or an open circuit).

REMARK 3.6.– Calibration kits including short-circuits, open circuits and matched loads (commonly referred to as "50 Ω plugs") are commercially available (see Figure 3.5). This equipment is costly, despite its apparent simplicity, with prices ranging from a few hundred to a few thousand euros. These devices are essentially used in addition to a network analyzer to carry out calibration using the standard "SOLT" (Short, Open, Load, Thru) method as close as possible to the equipment under test (and to compensate the effect of the cables connecting the equipment to the analyzer: the extremities of the cables are assimilated to reference planes for the ports of the quadripole in question).

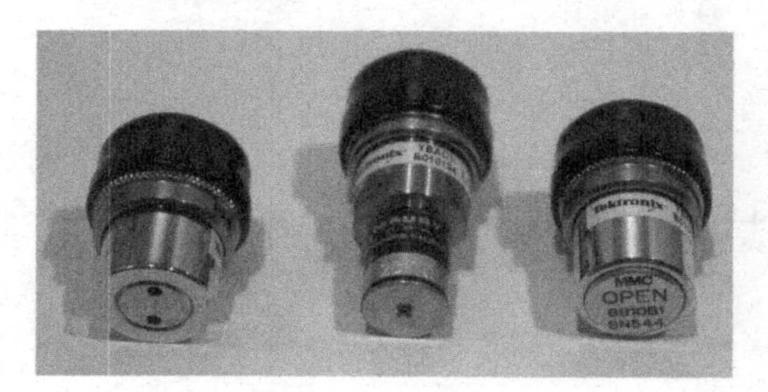


**Figure 3.5.** *Calibration kit for network analyzers*

3.2.5.3. *Network analyzers*

Network analyzers (see Figure 3.6) are sophisticated (and very expensive) devices which offer a range of functions, including certain aspects of impedance analysis. Network analyzers are able to fully characterize a quadripole (see Figure 3.7) using a matrix of four complex coefficients.

In electronics, the following matrices are generally used to describe quadripoles:

– impedance $(Z)$, linking the voltage vector $(V) = (\underline{v}_1, \underline{v}_2)^t$ to the current vector $(I) = (\underline{i}_1, \underline{i}_2)^t$;

– admittance $(Y)$, linking the two vectors $(V)$ and $(I)$ in the opposite direction;

– transfer $(T)$,[4] linking a hybrid "voltage/current" vector $(X_1) = (\underline{v}_1, \underline{i}_1)^t$ associated with port 1 (input) to a second vector $(X_2) = (\underline{v}_2, -\underline{i}_2)^t$ associated with port 2 (output).

**Figure 3.6.** *Agilent N5245A network analyzer (10 MHz–50 GHz)*

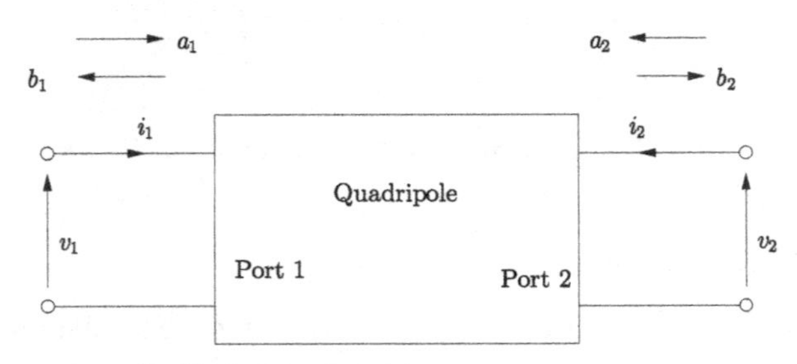


**Figure 3.7.** *Generic quadripole*

4 Also denoted as ABCD in certain publications.

The equations linking these different vectors and matrices are:

$$
\begin{aligned}
(V) &= \underbrace{\begin{pmatrix} \underline{Z}_{11} \; \underline{Z}_{12} \\ \underline{Z}_{21} \; \underline{Z}_{22} \end{pmatrix}}_{(Z)} . (I) \\
(I) &= \underbrace{\begin{pmatrix} \underline{Y}_{11} \; \underline{Y}_{12} \\ \underline{Y}_{21} \; \underline{Y}_{22} \end{pmatrix}}_{(Y)} . (V) \\
(X_1) &= \underbrace{\begin{pmatrix} \underline{T}_{11} \; \underline{T}_{12} \\ \underline{T}_{21} \; \underline{T}_{22} \end{pmatrix}}_{(T)} . (X_2)
\end{aligned}
\qquad [3.80]
$$

In the RF domain, a representation using "S" parameters is preferred – $\underline{S}_{ij}$ (scattering parameters) – which establishes a link between the incident waves $a_i$ and reflected waves $b_i$ at ports 1 and 2 ($i = 1$ or 2) of the quadripole:

$$
\begin{pmatrix} b_1 \\ b_2 \end{pmatrix} = \underbrace{\begin{pmatrix} \underline{S}_{11} \; \underline{S}_{12} \\ \underline{S}_{21} \; \underline{S}_{22} \end{pmatrix}}_{(S)} . \begin{pmatrix} a_1 \\ a_2 \end{pmatrix} \qquad [3.81]
$$

These are complex coefficients; when the device is able to fully determine the values of the $\underline{S}_{ij}$ coefficients, the term vector network analyzer (VNA) is used. A device which is only able to evaluate the moduli of these coefficients is known as a scalar analyzer.

REMARK 3.7.– This type of device is only really useful when considering propagation phenomena in circuits with distributed components; in these cases, voltage and current propagation produces visible effects (standing-waves in mismatched circuits).

Note that this equipment is particularly useful when considering variable signals with very high frequencies in relation to the propagation speed of electromagnetic waves and the size of the circuits involved. In this context, network analyzers are especially valuable. For illustrative purposes, the physical meanings of the coefficients $\underline{S}_{ij}$ of matrix $(S)$ may be specified:

– coefficient $\underline{S}_{11}$ (respectively $\underline{S}_{22}$) is a translation of the reflection phenomenon of the incident wave $a_1$ (respectively $a_2$) at port 1 (respectively 2);

– coefficient $\underline{S}_{12}$ (respectively $\underline{S}_{21}$) is a translation of the transfer phenomenon of the incident wave $a_1$ (respectively $a_2$) toward port 2 (respectively 1).

Waves $a_i$ and $b_i$ are specific quantities (neither voltages, currents, nor powers) used very specifically in the RF domains, and are square roots of powers, in that, for example, $a_1$ (the incident wave at port 1) is obtained from the incidence voltage (at the same port) using the formula:

$$a_1 = \frac{V_{i1}}{\sqrt{Z_0}} \qquad [3.82]$$

where $Z_0$ is the characteristic impedance (generally $50\,\Omega$) of the cables used between the analyzer and the device under test (DUT).

### 3.2.5.4. *Transfer relationships*

Although matrix $(S)$ supplies as much information related to a quadripole as matrices $(Z)$, $(Y)$ and $(T)$, it should be noted that this matrix cannot be "cascaded", unlike the transfer matrix $(T)$, and does not allow calculation of the parameters of series or parallel connections of quadripoles, something which is possible using matrices $(Y)$ and $(Z)$, respectively. The transfer relationships between these different representations are therefore particularly useful. A summary of the required relationships, taken from [FRI 94], is presented in Tables 3.1, 3.2 and 3.3. The impedance connected to port 1 of the quadripole is denoted as $Z_{01}$, and $Z_{02}$ is the impedance connected to port 2 of the same quadripole. Let $R_{01} = \mathfrak{Re}\,[Z_{01}]$ and $R_{02} = \mathfrak{Re}\,[Z_{02}]$.

| Passage from $(S)$ to $(Z)$ | |
|---|---|
| $S_{11} = \frac{(Z_{11}-Z_{01}^*)(Z_{22}+Z_{02})-Z_{12}Z_{21}}{(Z_{11}+Z_{01})(Z_{22}+Z_{02})-Z_{12}Z_{21}}$ | $S_{12} = \frac{2Z_{12}\sqrt{R_{01}R_{02}}}{(Z_{11}+Z_{01})(Z_{22}+Z_{02})-Z_{12}Z_{21}}$ |
| $S_{21} = \frac{2Z_{21}\sqrt{R_{01}R_{02}}}{(Z_{11}+Z_{01})(Z_{22}+Z_{02})-Z_{12}Z_{21}}$ | $S_{22} = \frac{(Z_{11}+Z_{01})(Z_{22}-Z_{02}^*)-Z_{12}Z_{21}}{(Z_{11}+Z_{01})(Z_{22}+Z_{02})-Z_{12}Z_{21}}$ |
| Passage from $(Z)$ to $(S)$ | |
| – | – |
| – | – |

**Table 3.1.** *Transfer relationships between matrices $(S)$ and $(Z)$*

REMARK 3.8.– No transfer relationship is indicated for a passage from matrix $(Z)$ to matrix $(Z)$, as in Dean Frickey's original article. Note that this transformation can be carried out *via* the admittance matrix $(Y) = (Z)^{-1}$.

Transfer relationships allow us to use the properties of matrices $(Z)$, $(Y)$ and $(T)$ applied to connections (series, parallel or cascaded) of quadripoles, summarized in Figure 3.8. Noting $(Z_1)$, $(Y_1)$ and $(T_1)$, the matrices linked to quadripole 1 and $(Z_2)$, $(Y_2)$ and $(T_2)$, the matrices linked to quadripole 2, we obtain the global matrices $(Z_g)$, $(Y_g)$ and $(T_g)$ of each connection verifying the following relationships:

– for a series connection $(Z_g) = (Z_1) + (Z_2)$;

– for a parallel connection $(Y_g) = (Y_1) + (Y_2)$;

– for a cascade connection $(T_g) = (T_1) + (T_2)$.

| Passage from $(S)$ to $(Y)$ | |
|---|---|
| $S_{11} = \dfrac{(1 - Y_{11} Z_{01}^*)(1 + Y_{22} Z_{02}) - Y_{12} Y_{21} Z_{01}^* Z_{02}}{(1 + Y_{11} Z_{01})(1 + Y_{22} Z_{02}) - Y_{12} Y_{21} Z_{01} Z_{02}}$ | $S_{12} = \dfrac{-2Y_{12}\sqrt{R_{01} R_{02}}}{(1 + Y_{11} Z_{01})(1 + Y_{22} Z_{02}) - Y_{12} Y_{21} Z_{01} Z_{02}}$ |
| $S_{21} = \dfrac{-2Y_{21}\sqrt{R_{01} R_{02}}}{(1 + Y_{11} Z_{01})(1 + Y_{22} Z_{02}) - Y_{12} Y_{21} Z_{01} Z_{02}}$ | $S_{22} = \dfrac{(1 + Y_{11} Z_{01})(1 - Y_{22} Z_{02}^*) - Y_{12} Y_{21} Z_{01} Z_{02}^*}{(1 + Y_{11} Z_{01})(1 + Y_{22} Z_{02}) - Y_{12} Y_{21} Z_{01} Z_{02}}$ |
| **Passage from $(Y)$ to $(S)$** | |
| $Y_{11} = \dfrac{(1 - S_{11})(Z_{02}^* + S_{22} Z_{02}) + S_{12} S_{21} Z_{02}}{(Z_{01}^* - S_{11} Z_{01})(Z_{02}^* + S_{22} Z_{02}) - S_{12} S_{21} Z_{01} Z_{02}}$ | $Y_{12} = \dfrac{-2S_{12}\sqrt{R_{01} R_{02}}}{(Z_{01}^* - S_{11} Z_{01})(Z_{02}^* + S_{22} Z_{02}) - S_{12} S_{21} Z_{01} Z_{02}}$ |
| $Y_{21} = \dfrac{-2S_{21}\sqrt{R_{01} R_{02}}}{(Z_{01}^* - S_{11} Z_{01})(Z_{02}^* + S_{22} Z_{02}) - S_{12} S_{21} Z_{01} Z_{02}}$ | $Y_{22} = \dfrac{(1 - S_{22})(Z_{01}^* + S_{11} Z_{01}) + S_{12} S_{21} Z_{01}}{(Z_{01}^* - S_{11} Z_{01})(Z_{02}^* + S_{22} Z_{02}) - S_{12} S_{21} Z_{01} Z_{02}}$ |

**Table 3.2.** *Transfer relationships between matrices $(S)$ and $(Y)$*

| Passage from $(S)$ to $(T)$ | |
|---|---|
| $S_{11} = \frac{T_{11}Z_{02}+T_{12}-T_{21}Z_{01}^{*}Z_{02}-T_{22}Z_{01}^{*}}{T_{11}Z_{02}+T_{12}+T_{21}Z_{01}^{*}Z_{02}+T_{22}Z_{01}}$ | $S_{12} = \frac{2(T_{11}T_{22}-T_{12}T_{21})\sqrt{R_{01}R_{02}}}{T_{11}Z_{02}+T_{12}+T_{21}Z_{01}^{*}Z_{02}+T_{22}Z_{01}}$ |
| $S_{21} = \frac{2\sqrt{R_{01}R_{02}}}{T_{11}Z_{02}+T_{12}+T_{21}Z_{01}^{*}Z_{02}+T_{22}Z_{01}}$ | $S_{22} = \frac{-T_{11}Z_{02}^{*}+T_{12}-T_{21}Z_{01}Z_{02}^{*}+T_{22}Z_{01}}{T_{11}Z_{02}+T_{12}+T_{21}Z_{01}^{*}Z_{02}+T_{22}Z_{01}}$ |
| **Passage from $(T)$ to $(S)$** | |
| $T_{11} = \frac{(1-S_{22})(Z_{01}^{*}+S_{11}Z_{01})+S_{12}S_{21}Z_{01}}{2S_{21}\sqrt{R_{01}R_{02}}}$ | $T_{12} = \frac{(Z_{01}^{*}+S_{11}Z_{01})(Z_{02}^{*}+S_{22}Z_{02})-S_{12}S_{21}Z_{01}Z_{02}}{2S_{21}\sqrt{R_{01}R_{02}}}$ |
| $T_{21} = \frac{(1-S_{11})(1-S_{22})+S_{12}S_{21}}{2S_{21}\sqrt{R_{01}R_{02}}}$ | $T_{22} = \frac{(1-S_{11})(Z_{02}^{*}+S_{22}Z_{02})+S_{12}S_{21}Z_{02}}{2S_{21}\sqrt{R_{01}R_{02}}}$ |

**Table 3.3.** *Transfer relationships between matrices $(S)$ and $(T)$*

### 3.2.6. *Transient mode*

#### 3.2.6.1. *Context of power electronics*

The study of transmission lines presented above applied to operations in sinusoidal steady-state. However, in the context of power electronics, while periodic modes do exist, these are generally based on stepped waveforms. It is therefore useful to study the behavior of a transmission line, whether matched or mismatched to the source and load, in a transient context of the "step response" type.

While analytical calculations are possible (mobile wave method), they are fastidious after a certain number of iterations, and can result in errors when carried out manually. A graphical analysis tool has therefore been developed for this purpose, known as a Bergeron diagram. This will be discussed in the next section.

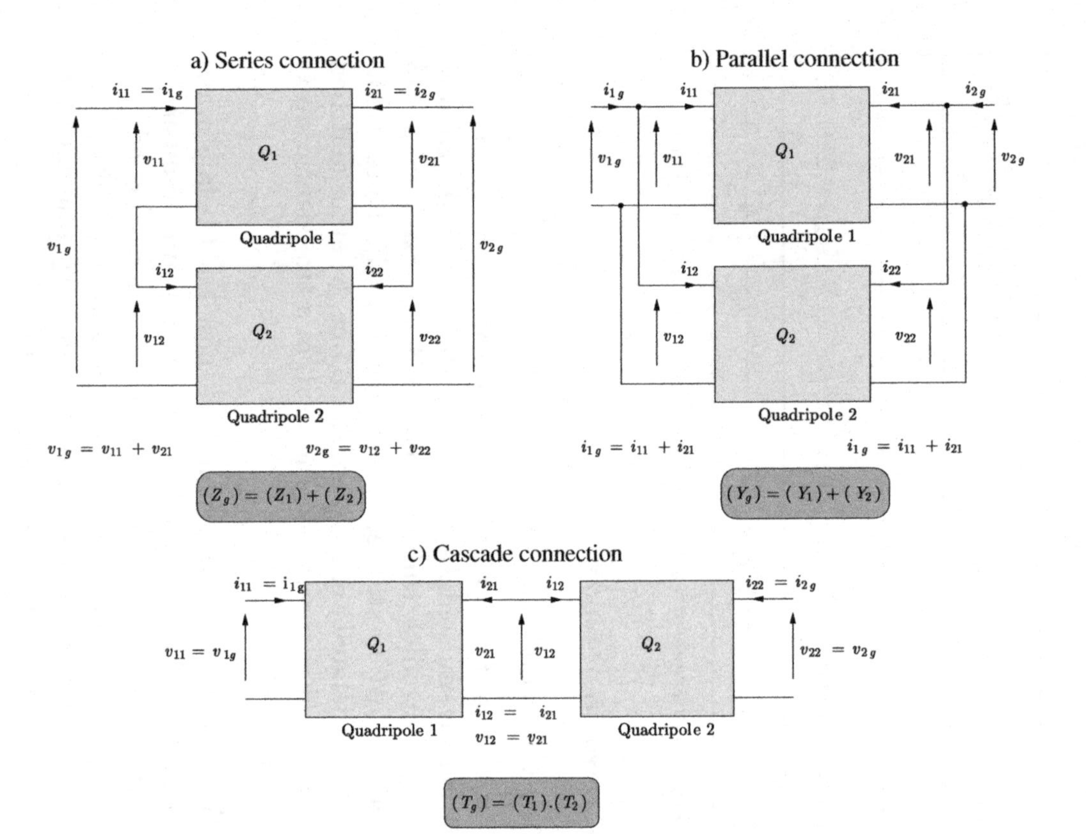


**Figure 3.8.** *Quadripole connections*

3.2.6.2. *Reflections in open or short circuits*

When a step is generated at $t = 0$ by a source (with open voltage amplitude $E_0$) matched to a cable, the voltage which is instantaneously visible at the cable input point will be equal to $E_0/2$. This voltage wave then propagates along the length of the cable, of length $L_c$, at a characteristic speed $c'$. The time taken by the wave to travel from one end of the cable to the other is denoted as $\tau$.

In the case of a circuit which is open at the "load side" extremity, the wave is reflected with the same amplitude and polarity as the incident wave. The amplitude of the input voltage is therefore $E_0$ after a time $t = 2\tau$.

In the case of a short-circuit, the wave is still reflected with the same amplitude, but the polarity is reversed in relation to the incident wave. The amplitude of the input voltage therefore falls to 0 after the same period of time $t = 2\tau$.

These two examples are illustrated in Figure 3.9.

3.2.6.3. *Bergeron diagrams*

The Bergeron diagram consists of tracing the evolution of a wave between the source and the load (by iterations) in a reference frame $(V, Z_c.I)$, where $Z_c$ is the characteristic impedance of the cable between the two, based on the characteristics of the source and the load.

As the characteristics of these two elements can only be static, this tool may only be applied if reactive elements (inductances or capacitances) are present. However, it is perfectly suitable for use with nonlinear elements (such as diodes).

An example of a trace is shown in Figure 3.10. This example shows the usage principle of the Bergeron diagram: displacements between the source characteristic and the load characteristic occur by segments with a line at ±45°, in accordance with the behavior of a cable of impedance $Z_c$ in the plane $(V, Z_C.I)$. Based on successive operating points on the source side and on the load side, the temporal waveforms of the corresponding voltages (Figure 3.11) can easily be found; these waveforms evolve by steps of duration $2\tau$, with $\tau = \frac{2L_c}{c'}$, where $L_C$ is the length of the cable and $c'$ is the propagation speed of the wave (this is therefore the time taken for the wave to travel from one end to the other and back again).

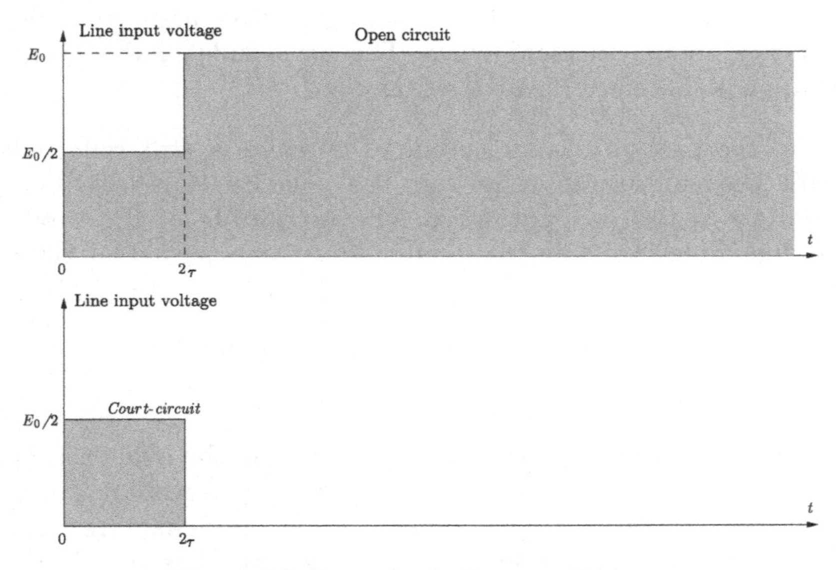


**Figure 3.9.** *Example of a Bergeron diagram*

In this example, a succession of operating points are denoted as A, B, C, D, ...∞. These points alternatively represent the state of the extremity on the source side (A at $t = 0$, C at instant $2\tau$) and of the extremity on the load side (B at instant $t = \tau$, D at instant $t = 3\tau$). The system finally converges (after an infinite number of iterations in theory, but relatively quickly in practice) toward the point of intersection between the source and load characteristics (this point is denoted as $\infty$ in the figure). This behavior is coherent with our classic experiment (at low frequency level) concerning a coaxial cable connection, which is ideally assimilated to two equipotentials.

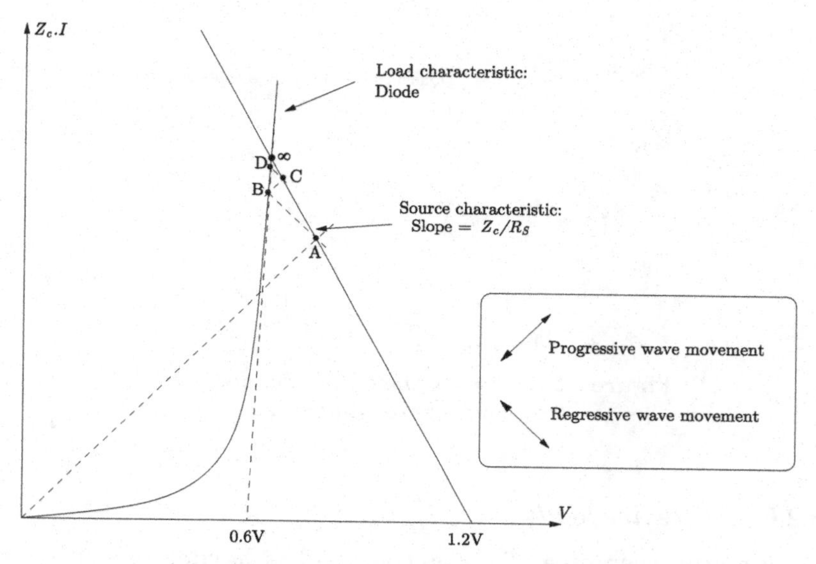


**Figure 3.10.** *Example of a Bergeron diagram*

### 3.2.6.4. *Time domain reflectometry*

Time domain reflectometry involves analyzing the response to a step (or a pulse) in a cable of known impedance for diagnostic purposes. When the cable is connected to another cable, variations in impedance are frequent, and reflections generally occur when impedance variations exist. By measuring the delays and the amplitude of “accidents” observed in the cable input voltage, information can be obtained regarding the configuration of the cable, any faults (short-circuits and discontinuities) and their location. Devices designed for this purpose are known as Time domain reflectometers (TDRs). This type of equipment exists not only for copper cables (Metallic TDR), but also for optical fibers (Optical TDR) – see Figure 3.12 (left and right, respectively).

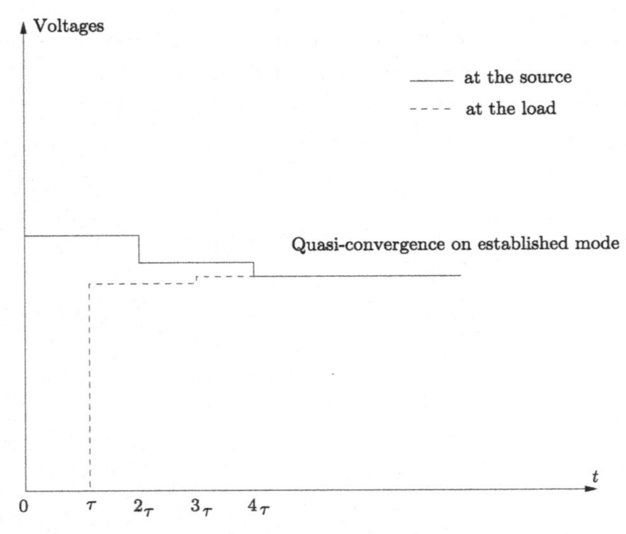


**Figure 3.11.** *Chronogram of voltages deduced from the Bergeron diagram*

### 3.2.6.5. *Reactive loads*

The step response of a reactive load is harder to study, as Bergson diagrams, designed for static situations, can no longer be used. For this reason, simulation is generally preferred. The "classic" behavior of an $RL$ or $RC$ circuit should, however, be kept in mind in order to verify the coherence of the obtained result with usual "low frequency" behaviors. For illustrative purposes, consider the case of a voltage source (step $E_0$) with a series impedance $R_S$ of $50\,\Omega$ (resistive) connected to a cable with a characteristic impedance of $50\,\Omega$, the extremity of which is connected to a capacitor of capacitance $C = 1\,\mu\text{F}$.

Classically (leaving aside the cable), the system made up of the impedance source and the capacitor allows the generator output voltage $v(t)$[5] to be written in the form:

$$v(t) = E_0 . \left(1 - \mathrm{e}^{-t/\tau'}\right) \text{ with } \tau' = R_S C = 50\,\mu\text{s} \qquad [3.83]$$

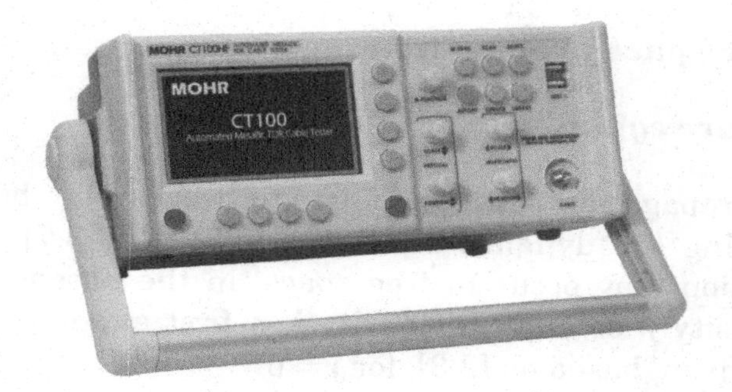


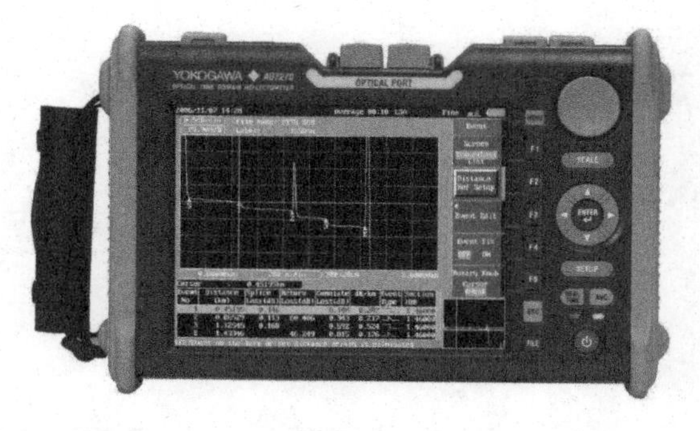


**Figure 3.12.** *Time domain reflectometers (TDRs) for "copper" cables (left) and optical fibers (right)*

In practice, at $t = 0$, the voltage wave applied to the cable perceives the characteristic impedance of the cable. In these conditions, a divider bridge with a gain of 1/2 is obtained, with a voltage source $E_0$ producing a voltage $E_0/2$. This wave will remain unchanged until it has traveled through the cable in both directions (outward journey of duration $\tau$ and inward journey, after time-variable reflection, of duration $\tau$). In fact,

5 Presumed to be equal to the voltage at the capacitor terminals in LF, as the coaxial cable is assimilated to two equipotentials.

the expected waveform in LF, expressed in equation [3.83], is preceded by a pulse of constant amplitude $E_0/2$ and duration $2\tau$, as shown in the time curve in Figure 3.13.

## 3.3. Free-space propagation

### 3.3.1. *Wave equation*

The propagation of electromagnetic waves may be shown by coupling the dynamic equations [3.3] and [3.7]. As this propagation may occur in free space, in the absence of any load density $\rho$ or current density j, a first expression of $\frac{\partial \mathbf{E}}{\partial t}$ may be given, based on [3.3], for $\mathbf{j} = 0$:

$$\frac{\partial \mathbf{E}}{\partial t} = \frac{1}{\mu_0 \varepsilon_0} \mathbf{curl}\,\mathbf{B} \qquad [3.84]$$

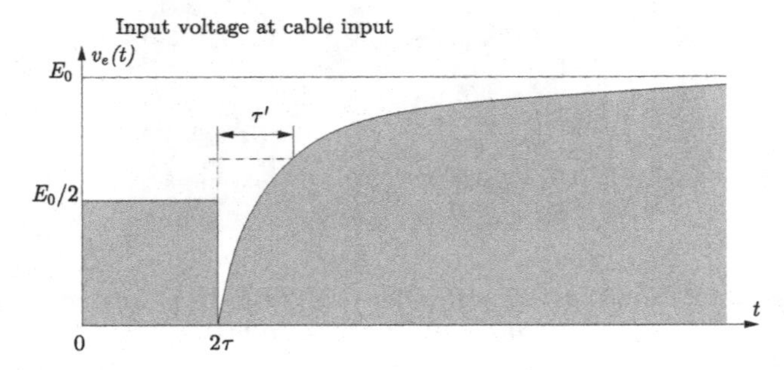


**Figure 3.13.** *Step response of a "matched cable + capacitor " system*

If the two sides of [3.7] are derived in relation to time, this result may be replaced in the obtained equation:

$$\frac{1}{\mu_0 \varepsilon_0} \cdot \mathbf{curl}\,\mathbf{curl}\,\mathbf{B} = -\frac{\partial^2 \mathbf{B}}{\partial t^2} \qquad [3.85]$$

The notion of the vector laplacian should then be introduced, defined, in the Cartesian framework, as:

$$\boldsymbol{\Delta}\,\mathbf{B} = \begin{pmatrix} \frac{\partial^2 B_x}{\partial x^2} + \frac{\partial^2 B_x}{\partial y^2} + \frac{\partial^2 B_x}{\partial z^2} \\ \frac{\partial^2 B_y}{\partial x^2} + \frac{\partial^2 B_y}{\partial y^2} + \frac{\partial^2 B_y}{\partial z^2} \\ \frac{\partial^2 B_z}{\partial x^2} + \frac{\partial^2 B_z}{\partial y^2} + \frac{\partial^2 B_z}{\partial z^2} \end{pmatrix} \qquad [3.86]$$

where $B_x$, $B_y$ and $B_z$ are the components of vector $\mathbf{B}$ in the Cartesian reference frame under consideration.

A notable equality allows us to write:

$$\boldsymbol{\Delta}\,\mathbf{B} = \mathbf{grad}\,\mathrm{div}\,\mathbf{B} - \mathbf{curl}\,\mathbf{curl}\,\mathbf{B} \qquad [3.87]$$

Moreover, based on the magnetic flux conservation equation [3.6], this equation may be reduced to:

$$\boldsymbol{\Delta}\,\mathbf{B} = -\mathbf{curl}\,\mathbf{curl}\,\mathbf{B} \qquad [3.88]$$

Using this result, [3.85] can be rewritten as:

$$\boldsymbol{\Delta}\,\mathbf{B} - \mu_0\varepsilon_0\frac{\partial^2\mathbf{B}}{\partial t^2} = \mathbf{0} \qquad [3.89]$$

REMARK 3.9.– The second side is a zero vector, denoted as $\mathbf{0}$, and not the scalar $0$.

This 2nd-order partially derived equation is known as the wave equation, where coefficient $\mu_0\varepsilon_0$ is homogeneous to the inverse square of the propagation speeds of the waves constituting solutions to the equation. This speed is denoted as $c$, and is thus linked to the electromagnetic properties of free space by the relationship:

$$\mu_0\varepsilon_0 c^2 = 1 \qquad [3.90]$$

This speed is known as celerity, and is the propagation speed of electromagnetic waves in a vacuum, i.e. the speed of light, approximately $3 \times 10^8$ m/s.

A single-dimension case, such that $B_y = B_z = 0$, is easier to analyze. In these conditions, the induction field **B** becomes a scalar $B_x$, and equation [3.89] becomes:

$$\frac{\partial^2 B_x}{\partial x^2} - \frac{1}{c^2} \cdot \frac{\partial^2 B_x}{\partial t^2} = 0 \qquad [3.91]$$

It is easy to verify that the solutions to this equation are of the form:

$$B_x(x,t) = f^+(x - ct) + f^-(x + ct) \qquad [3.92]$$

where functions $f^+(\cdot)$ and $f^-(\cdot)$ are waves which propagate in opposite directions. In the case of guided propagation, with a specified preferred direction of propagation (in this case, the positive direction), the terms "incident wave" and "reflected wave" are used.

Equation [3.89] can be clearly rewritten using field **H**, as these two quantities are proportional.

The wave equation relating to the electric field is established by applying the same approach used in equation [3.85]. Instead of deriving this equation in relation to time, we derive equation [3.3], still using the hypothesis of a current density j of zero:

$$\mathbf{curl}\,\frac{\partial \mathbf{H}}{\partial t} = \frac{\partial^2 \mathbf{D}}{\partial t^2} \qquad [3.93]$$

First, **H** and **D** are replaced by their respective expressions as functions of **B** and **E**:

$$\frac{1}{\mu_0}\mathbf{curl}\,\frac{\partial \mathbf{B}}{\partial t} = \varepsilon_0 \frac{\partial^2 \mathbf{E}}{\partial t^2} \qquad [3.94]$$

$\frac{\partial \mathbf{B}}{\partial t}$ is then replaced by $-\text{rot}\,\mathbf{E}$ following [3.17]. Formula [3.87] is then applied to field **E**, giving:

$$-\textbf{grad}\,\text{div}\,\mathbf{E} + \mathbf{\Delta}\,\mathbf{E} = \mu_0\varepsilon_0\frac{\partial^2\mathbf{E}}{\partial t^2} \qquad [3.95]$$

In this case, the term $\textbf{grad}\,\text{div}\,\mathbf{E}$ has not been yet eliminated. Note, however, that in the absence of a load density $\rho$, $\text{div}\,\mathbf{E} = 0$. Consequently, the previous equation can be reduced:

$$\mathbf{\Delta}\,\mathbf{E} - \mu_0\varepsilon_0\frac{\partial^2\mathbf{E}}{\partial t^2} = \mathbf{0} \qquad [3.96]$$

This results in the same equation for **B** as for **H**. This equation can evidently also be rewritten using field **D**, for the same reason as in the case of the magnetic field.

### 3.3.2. *Speed, frequency and wavelength*

In the previous section, electromagnetic equations were used to show that electrical and magnetic fields propagate through a vacuum or through air at the speed of light $c$ (in the broadest possible sense, as this property applies to **E**, **D**, **B** and **H**). Consequently, when these quantities are periodic (of period $T = 1/f$) as a function of time at a point in space, this periodicity is also found at spatial level; the wavelength $\lambda$ is thus the spatial equivalent of the period in the time domain:

$$\lambda = \frac{c}{f} \qquad [3.97]$$

This quantity is particularly important when considering the validity of a "circuit" type modeling of an electrical or electronic system.

In circuit theory, wires are assimilated to equipotentials, and the current is considered to be the same at any point in a circuit loop. This is not the case in practice in HF mode, in circuits where the characteristic dimensions are not negligible in relation to the wavelength: in these conditions, a distributed constant modeling is required in order to take account of propagation phenomena.

### 3.3.3. *Wave impedance*

The notion of wave impedance (more precisely the intrinsic impedance of a vacuum, in this specific case) may be used to establish a link between field **E** and field **H** of an electromagnetic wave. This is an important parameter in wave propagation. It is easy to verify that, from a dimensional perspective, the ratio **E/H** is equivalent to an impedance (in Ohms), in that **E** is expressed in V/m, while **H** is expressed in A/m. The simplest propagation mode to study is that of permanent steady-state (with angular frequency $\omega$), with two fields **E** and **H** evolving in a plane perpendicular to the direction of propagation. This is known as transverse electromagnetic (TEM) propagation.

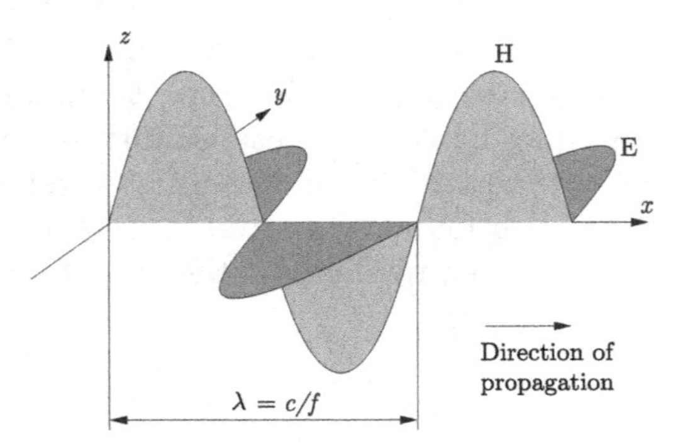


**Figure 3.14.** *Propagation of a TEM wave along the $x$ axis*

As an example, we will consider the simple case (see Figure 3.14) of propagation of a progressive (TEM) wave with two orthogonal[6] fields **E** and **H** (one directed along the $y$ axis and the other following the $z$ axis – the specific case of a wave with rectilinear polarization) along the $x$ axis.

Take:

$$\begin{cases} \mathbf{E}(x,t) = E(x,t).\mathbf{e}_y \\ \mathbf{H}(x,t) = H(x,t).\mathbf{e}_z \end{cases} \quad [3.98]$$

where $\mathbf{e}_y$ and $\mathbf{e}_z$ are the direction vectors corresponding to the $y$ and $z$ axes, respectively ($\mathbf{e}_x$ is the equivalent vector for the $x$ axis).

For calculation purposes, the real moduli $E(x,t)$ and $H(x,t)$ of the fields will be replaced by equivalent complex amplitudes ($\underline{E}(x,t)$ and $\underline{H}(x,t)$ respectively), expressed as:

$$\underline{E}(x,t) = E_0.\mathrm{e}^{j(\omega t - kx)} \quad [3.99]$$

$$\underline{H}(x,t) = H_0.\mathrm{e}^{j(\omega t - kx + \phi)} \quad [3.100]$$

where $k = \frac{2\pi}{\lambda}$ is the wavenumber.

We then need to verify the conditions $E_0$, $H_0$ and $\phi$ in which Maxwell's equations are verified. Specifically, the validity of the Maxwell–Faraday equation may be tested:

$$\mathbf{curl}\,\mathbf{E} = -\mu_0 \frac{\partial \mathbf{H}}{\partial t} \quad [3.101]$$

6 This is a generic property of plane waves.

The rotational formula is then applied using Cartesian coordinates (introducing the Nabla operator $\nabla$):

$$\mathbf{curl}\,\mathbf{E} = \nabla \times \mathbf{E} = \begin{pmatrix} \frac{\partial}{\partial x} \\ \frac{\partial}{\partial y} \\ \frac{\partial}{\partial z} \end{pmatrix} \times \begin{pmatrix} E_x \\ E_y \\ E_z \end{pmatrix} = \begin{pmatrix} \frac{\partial E_z}{\partial y} - \frac{\partial E_y}{\partial z} \\ \frac{\partial E_x}{\partial z} - \frac{\partial E_z}{\partial x} \\ \frac{\partial E_y}{\partial x} - \frac{\partial E_x}{\partial y} \end{pmatrix} \quad [3.102]$$

In this case, where $E_x = E_z = 0$ (and $E_y$ is only dependent on $x$), this rotational can be simplified as follows:

$$\mathbf{curl}\,\mathbf{E} = \begin{pmatrix} 0 \\ 0 \\ \frac{\partial E_y}{\partial x} \end{pmatrix} = -jkE_0 \begin{pmatrix} 0 \\ 0 \\ \mathrm{e}^{j(\omega t - kx)} \end{pmatrix} \quad [3.103]$$

This rotation may then be identified to the right hand side of equation [3.101], i.e.:

$$-\mu_0 \frac{\partial \mathbf{H}}{\partial t} = -j\mu_0 \omega H_0 \begin{pmatrix} 0 \\ 0 \\ \mathrm{e}^{j(\omega t - kx + \phi)} \end{pmatrix} \quad [3.104]$$

The $z$ axis components of the two sides should then be identified in order to obtain the desired result:

$$kE_0 = \mu_0 \omega H_0 \mathrm{e}^{j\phi} \quad [3.105]$$

The moduli and arguments of the two sides can then be identified in order to establish the following relationships:

$$\frac{E_0}{H_0} = \frac{\mu_0 \omega}{k} = \sqrt{\frac{\mu_0}{\varepsilon_0}} \quad [3.106]$$

and:

$$\phi = 0 \qquad [3.107]$$

Let $Z_0$ be the ratio $E_0/H_0$, known as the intrinsic impedance of a vacuum, with an approximate value of $120\pi \simeq 376.8\,\Omega$.

This result allows a link to be established between the electric field and the magnetic field emitted by a distant antenna. This is known as far field behavior (at a distance $D$ from the antenna which is high in relation to the characteristic dimensions). This context considerably simplifies study of the radiation of an antenna, in that the emitted field tends systematically toward a plane wave for any type of antenna. Unfortunately, near-field study is more relevant in the case of an EMC study, as:

– the polluting device is not an efficient emitter, and its range is low;

– victims are generally close to the radiating element (and may, moreover, be part of the same equipment).

Radiating sources therefore often need to be characterized in a near-field, requiring field calculations which may be complex. The equations established above showed the capacity of electrical and magnetic fields to propagate through space, but the lack of a source ($\rho$ or j) prevented true demonstration of the behavior of radiating elements. This point will be discussed in the following section.

### 3.3.4. *The Biot–Savart law*

The Biot–Savart law is used to calculate an infinitesimal magnetic filed $d\mathbf{B}$ (r) produced at a point localized by vector r by a circuit element of length $d\mathrm{l}$, situated at the point defined by $\mathbf{r}'$, in which a current $I$ circulated:

$$d\mathbf{B}(\mathbf{r}) = \frac{\mu_0}{4\pi} \cdot \frac{I d\mathbf{l} \times (\mathbf{r} - \mathbf{r}')}{\|\mathbf{r} - \mathbf{r}'\|^3} \qquad [3.108]$$

REMARK 3.10.– This law will not be demonstrated here. Note, however, that it is derived from Maxwell's equations and can easily be obtained when considering a circuit element which is quasi-punctual (very short, and made up of a wire with a negligible cross-section) in relation to the point $\mathbf{r}$ at which the field is calculated. Moreover, the notion of vector potential $\mathbf{A}$ is generally used in the demonstration (although this is not strictly necessary), such that $\mathbf{B} = \mathbf{rot}\,\mathbf{A}$, as we know that $\mathrm{div}\,(\mathbf{rot}\,\bullet) = 0$ (this guarantees that $\mathbf{B}$ will satisfy the magnetic flux conservation equation). The equation in terms of $\mathbf{A}$ takes a form very similar to equation [3.108]:

$$d\mathbf{A}(\mathbf{r}) = \frac{\mu_0}{4\pi} \cdot \frac{\mathbf{j}.d\mathcal{V}(\mathbf{r}')}{\|\mathbf{r} - \mathbf{r}'\|} \qquad [3.109]$$

where $d\mathcal{V}(\mathbf{r})'$ is a volume element surrounding the considered point $\mathbf{r}'$ of the source.

When considering variable modes (and not the calculation of quasi-static fields), a propagation time needs to be taken into account for actions at a distance. These are known as delayed-action potentials. The duration of the delay is linked to the distance separating the source (a wire carrying a current at point $\mathbf{r}'$) from the point under consideration ($\mathbf{r}$) when calculating the field and the propagation speed ($c$). This results in the equation can be written as:

$$d\mathbf{A}(\mathbf{r}) = \frac{\mu_0}{4\pi} \cdot \frac{\mathbf{j}\left(\mathbf{r}', t - \frac{\|\mathbf{r}-\mathbf{r}'\|}{c}\right).d\mathcal{V}(\mathbf{r}')}{\|\mathbf{r} - \mathbf{r}'\|} \qquad [3.110]$$

An expression of the electrical (scalar) potential $V$ can also be given for any point in the space, taking account of the propagation phenomenon. The corresponding equation is:

$$dV(\mathbf{r}) = \frac{1}{4\pi\varepsilon_0} \cdot \frac{\rho\left(\mathbf{r}', t - \frac{\|\mathbf{r}-\mathbf{r}'\|}{c}\right).d\mathcal{V}(\mathbf{r}')}{\|\mathbf{r}-\mathbf{r}'\|} \quad [3.111]$$

### 3.3.5. *Emission principles*

An antenna is a conducting element designed to radiate an electromagnetic field into the surrounding space (antenna are also able to pick up radiation). The simplest radiating elements are electrical and magnetic dipoles. In this section, we will consider these elements, showing that these simple examples may be found in real-world cases of electromagnetic radiation within the context of power electronics.

#### 3.3.5.1. *Radiation of a dipolar antenna*

We will consider this antenna in the case of permanent steady-state and in a vacuum (or air). To simplify calculations, the equations of complex quantities will be expressed in the form $e^{j\omega t}$. First, consider the Maxwell–Ampère equation in this context:

$$\mathbf{curl\,H} = \mathbf{j} + j\omega\mathbf{D} = \mathbf{j} + j\omega\varepsilon_0\mathbf{E} \quad [3.112]$$

Calculating the divergence of this relationship, we obtain:

$$0 = \operatorname{div}\mathbf{j} + j\omega\rho \quad [3.113]$$

This is the electric field conservation equation, which allows a connection to be established between the volume density of the load $\rho$ and the current density $\mathbf{j}$:

$$\rho = -\frac{1}{j\omega}\operatorname{div}\mathbf{j} \quad [3.114]$$

In the case of a Hertz dipole (a conducting segment of length $L$ carrying a current $i_0$) observed from a large distance ($r \gg L$), the current density may be considered to be punctual (located at $x = y = z = 0$). Moreover, the $z$ axis is used as the axis of the dipole:

$$\mathbf{j} = i_0 L \delta(x) . \delta(y) . \delta(z) . \mathbf{e}_z \qquad [3.115]$$

where $\delta(\bullet)$ is the Dirac distribution.

In fact, note that current $i_0$ is actually a sinusoidal current, with the equivalent complex formulation:

$$i_0 = I_{max} . e^{j\omega t} \qquad [3.116]$$

The Dirac impulse may be derived in the direction of distribution in order to obtain the following boundary passage:

$$\frac{d\delta(z)}{dz} = \lim_{L \to 0} \frac{\delta\left(z + \frac{L}{2}\right) - \delta\left(z - \frac{L}{2}\right)}{L} \qquad [3.117]$$

Consequently, the load distribution $\rho$ linked to j for this dipole is:

$$\rho = -\lim_{L \to 0} \frac{i_0 \delta(x) \delta(y)}{j\omega} \left[ \delta\left(z + \frac{L}{2}\right) - \delta\left(z - \frac{L}{2}\right) \right] \qquad [3.118]$$

The fact that the source (segment carrying a current) is considered as a punctual element means that it can be assimilated to a dipole made up of two loads located at $z = -L/2$ and $+L/2$ (with $x = y = 0$).

Using this basis, the vector potential $\mathbf{A}(\mathbf{r})$ may be calculated at a point distant from the source by applying equation [3.110] to the delayed-action potential:

$$\mathbf{A}(\mathbf{r}) = \frac{\mu_0}{4\pi} \cdot \iiint \frac{\mathbf{j}\left(\mathbf{r}', t - \frac{\|\mathbf{r}-\mathbf{r}'\|}{c}\right).d\mathcal{V}(\mathbf{r}')}{\|\mathbf{r}-\mathbf{r}'\|}$$
$$= \frac{\mu_0 I_{max} L}{4\pi r} e^{j\left(\omega\left(t-\frac{r}{c}\right)\right)} \cdot \mathbf{e}_z \qquad [3.119]$$

where $r = \|\mathbf{r}\| \simeq \|\mathbf{r}-\mathbf{r}'\|$. This shows that the vector potential $\mathbf{A}(\mathbf{r})$ is colinear with the $z$ axis when the antenna is observed from a distance $r$ which is high in relation to the length $L$ of the antenna ($r \gg L$).

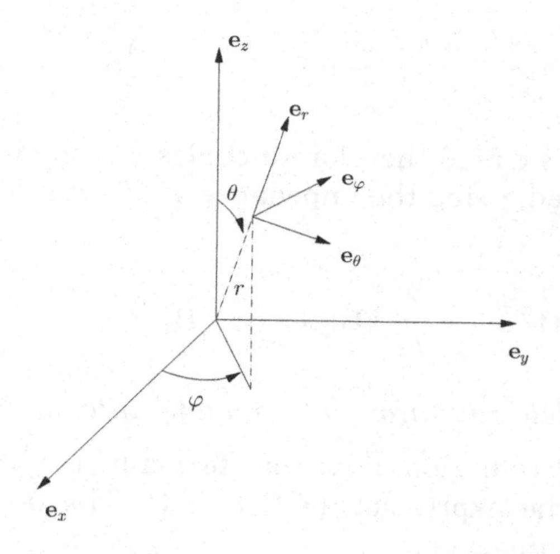


**Figure 3.15.** *Passage from Cartesian to spherical coordinates*

In a far-field situation, the problem is spherical. The representation of coordinate systems shown in Figure 3.15 is therefore used. Then, the induction should be calculated, based on the following property of the rotation of an axial field:

$$\mathbf{curl}\,(\mu.\mathbf{e}_z) = (\mathbf{grad}\,\mu) \times \mathbf{e}_z \qquad [3.120]$$

Hence, in this case:

$$\mathbf{B} = \mathbf{curl}\,\mathbf{A} = \frac{\mu_0 I_{max} L}{4\pi}\mathbf{grad}\left(\frac{\mathrm{e}^{j\left(\omega\left(t-\frac{r}{c}\right)\right)}}{r}\right) \times \mathbf{e}_z \qquad [3.121]$$

Using spherical coordinates, the gradient calculation gives:

$$\mathbf{grad}\left(\frac{\mathrm{e}^{j\omega\left(t-\frac{r}{c}\right)}}{r}\right) = -\frac{1}{r^2}\left(1+\frac{j\omega r}{c}\right)\mathrm{e}^{j\omega\left(t-\frac{r}{c}\right)}\mathbf{e}_r \qquad [3.122]$$

Thus, as $\mathbf{e}_r \times \mathbf{e}_z = -\sin\theta.\mathbf{e}_\varphi$:

$$\mathbf{B} = \frac{\mu_0 I_{max} L.\sin\theta}{4\pi}\left(\frac{1}{r^2}+\frac{j\omega}{rc}\right)\mathrm{e}^{j\omega\left(t-\frac{r}{c}\right)}.\mathbf{e}_\varphi \qquad [3.123]$$

The magnetic field therefore includes an azimuth. Field $\mathbf{E}$ is then deduced, using the impedance of a vacuum, as defined in [3.106]:

$$\mathbf{E} = Z_0\left\|\mathbf{H}\right\|.\mathbf{e}_\theta = \frac{Z_0}{\mu_0}\left\|\mathbf{B}\right\|.\mathbf{e}_\theta = c\left\|\mathbf{B}\right\|.\mathbf{e}_\theta \qquad [3.124]$$

3.3.5.2. *Far-field radiation diagram and antenna resistance*

In far-field radiation, note that terms in $1/r$ and $1/r^2$ are included in the expression of $\mathbf{E}$ and $\mathbf{B}$. The terms in $1/r^2$ become negligible when:

$$\frac{\omega}{cr} \gg \frac{1}{r^2} \qquad [3.125]$$

The notion of far-field is related to the wavelength $\lambda$ of the emitted signal. If the antenna is observed from a very long distance ($r \gg \lambda$ – typically $r > 10.\lambda$), the expressions of the two fields are simplified. For example, with induction $\mathbf{B}$:

$$\mathbf{B}(r,\theta) = j\frac{\mu_0 \omega I_{max} L.\sin\theta}{4\pi rc}\cdot\mathrm{e}^{j\omega\left(t-\frac{r}{c}\right)}.\mathbf{e}_\varphi \qquad [3.126]$$

There is a maximum emission for $\theta = \frac{\pi}{2}$, with an amplitude of:

$$B_{\max} = \frac{\mu_0 \omega I_{max} L}{4\pi r c} \qquad [3.127]$$

Although it is possible to create a trace of a radiation diagram in a magnetic induction field (or an electric field) for this dipolar antenna, it is generally better to use a "power flux" diagram. This is done using the Poynting vector $\mathfrak{P}$, defined as:

$$\mathfrak{P} = \mathbf{E} \times \mathbf{H} \qquad [3.128]$$

This vector is homogeneous to a surface power density (in W/m$^2$), so the flux of $\mathfrak{P}$ is simply calculated across a closed surface $\Sigma$ surrounding the emitting antenna in order to calculated the emitted power $P_{\text{em}}$:

$$P_{\text{em}} = \oiint_{\Sigma} \mathfrak{P} \cdot ds \qquad [3.129]$$

In this case, we obtain:

$$\|\mathfrak{P}\| = Z_0 . \left( \frac{\omega I_{max} L}{4\pi r c} \right)^2 \sin^2 \theta \qquad [3.130]$$

After normalization in relation to the maximum modulus of $\mathfrak{P}$ at a given distance $r$, an antenna radiation diagram is obtained in terms of "$\sin^2 \theta$". This diagram is shown in Figure 3.16. The antenna radiates around the $z$ axis in all directions, independently of $\varphi$; however, the intensity of the Poynting vector is dependent on $\theta$, with maximum emission at $\theta = \frac{\pi}{2}$ and zero radiation in the axis of the antenna. This is interesting from an EMC perspective, as when considering an emitting antenna for which emissions cannot be contained, this information allows us to determine which spatial zones will be subject to the highest and lowest (or even zero) interference. Similarly, if the antenna is likely to be subject to interference, the diagram shows the directions of highest and lowest sensitivity.

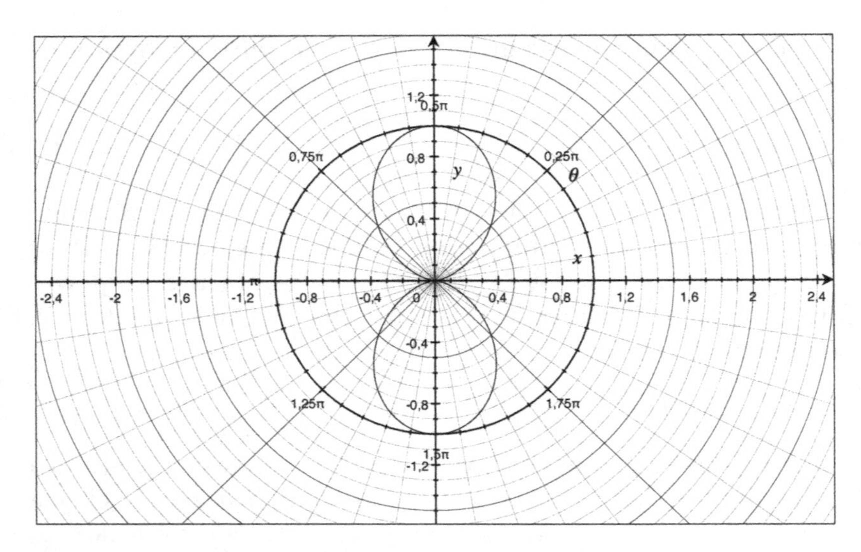


**Figure 3.16.** *Emission diagram for a dipolar antenna (the $x$ axis of the graph corresponds to the $z$ axis in the problem)*

REMARK 3.11.– Note that the power flux decreases in terms of $1/r^2$. This is coherent with an increasing distribution of the emitted power across the surface of concentric spheres (in $4\pi.r^2$) as the wave propagates (dilution of energy).

Based on this expression, which is dependent on the square of the current $i_0(t)$ circulating in the antenna, the equivalent resistance of the antenna $R_{\text{ant}}$ may be determined using the expression:

$$P_{\text{em}} = R_{\text{ant}}.I_{\text{RMS}}^2 \qquad [3.131]$$

where $I_{\text{RMS}}$ is the RMS value of the current.

In this example, the total emitted power $P_{\text{em}}$ may be evaluated by integrating the Poynting vector. This is not difficult, as it is radial:

$$\begin{aligned} P_{\text{em}} &= \int_{-\pi}^{+\pi} \int_{-\frac{\pi}{2}}^{+\frac{\pi}{2}} r^2 \left\| \mathfrak{P}(\theta) \right\| d\theta d\varphi = 2\pi \int_{-\frac{\pi}{2}}^{+\frac{\pi}{2}} r^2 \left\| \mathfrak{P}(\theta) \right\| d\theta \\ &= 2\pi Z_0 . \left( \frac{\omega I_{max} L}{4\pi c} \right)^2 \int_{-\frac{\pi}{2}}^{+\frac{\pi}{2}} \sin^2 \theta d\theta \end{aligned} \quad [3.132]$$

The charge $q$ multiplied by its displacement speed has already been seen to correspond to a current, and as $a_0\omega$ is the maximum charge displacement speed, the following expression may be used:

$$P_{\text{em}} = \frac{\pi I_{\text{max}}^2}{4} \sqrt{\frac{\mu_0}{\varepsilon_0}} \cdot \left( \frac{L}{\lambda} \right)^2 = \frac{\pi I_{\text{RMS}}^2}{2} \sqrt{\frac{\mu_0}{\varepsilon_0}} \cdot \left( \frac{L}{\lambda} \right)^2 \quad [3.133]$$

where $I_{\text{max}}$ and $I_{\text{RMS}}$ are, respectively, the amplitude and the RMS value of the current circulating in the antenna.

An equivalent antenna resistance can be identified in this expression:

$$R_{\text{ant}} = \frac{\pi}{2} \sqrt{\frac{\mu_0}{\varepsilon_0}} \cdot \left( \frac{L}{\lambda} \right)^2 \quad [3.134]$$

REMARK 3.12.– An antenna should not be completely reduced to the notion of impedance in relation to the power supply and to the radiation diagram (such as that shown in Figure 3.16). Even if two antenna are aligned in such a way that their respective maxima (of the emission/reception diagram) coincide, the receiving antenna will not necessarily pick up a signal transmitted by the emitting antenna. The polarizations also need to coincide in order for the excitation generated by the first antenna to be received, in the form of a signal, by the second antenna.

### 3.3.5.3. *Near-field radiation*

In the case of an electric dipole, the spherical components associated with the evaluation point of the electromagnetic field may be noted:

$$\begin{cases} H_\varphi = \frac{I_0 dl}{4\pi r^2} \sin\theta \cdot e^{j\omega\left(t-\frac{r}{c}\right)} \\ E_r = -jZ_0 \frac{I_0 dl \lambda}{4\pi^2 r^3} \cos\theta \cdot e^{j\omega\left(t-\frac{r}{c}\right)} \\ E_\theta = -jZ_0 \frac{I_0 dl \lambda}{8\pi^2 r^3} \cos\theta \cdot e^{j\omega\left(t-\frac{r}{c}\right)} \end{cases} \qquad [3.135]$$

The electromagnetic field is no longer perpendicular to the direction of propagation, and is therefore more complex than the far-field from a study perspective. Note, however, that in this configuration, the electric field dominates the magnetic field. In fact, a simple analysis of the ratio $\|\mathbf{E}\| / \|\mathbf{H}\|$ at all points in the space is sufficient to note that, with a far-field, the impedance is always the free-space impedance $Z_0$; in a near-field, however, the impedance is dependent on the nature of the antenna. In the case of an electric dipole, the electric field is dominant, while in the case of a magnetic dipole (i.e. a conducting coil), the magnetic field can be shown to dominate the electric field in the "near-field" approximation zone ($r \ll \frac{\lambda}{2\pi}$).

In simple cases, far-field radiation can be calculated analytically; however, near-field study is generally more difficult, and a digital approach is preferred for this type of problem. A wide range of commercial or freely-available software is available for this purpose. These programs are not necessarily designed for EMC, but are often intended for the study and design of antenna. One example is 4NEC2, a free program based on the method of moments, which uses the Numerical Electromagnetic Code (NEC) calculation code developed in 1981 by J. Burke and A. Poggio at the Lawrence Livermore Laboratory (USA) as part of a contract with the US Navy.

This software facilitates users to implant sources in an environment with 3D geometry. An example calculation is shown in Figure 3.17, with a representation of the intensity of the magnetic field generated by an antenna fixed on a simplified geometric model of a car. This mapping evidently shows that the field is reduced as the distance from the source (antenna) increases, but presents local maxima at certain points distant from the vehicle chassis. Note that this simulation is carried out using a finite zone of a horizontal section plane of the car, and the program has certain limitations in terms of simulable nodes (volume/surface of study, resolution).

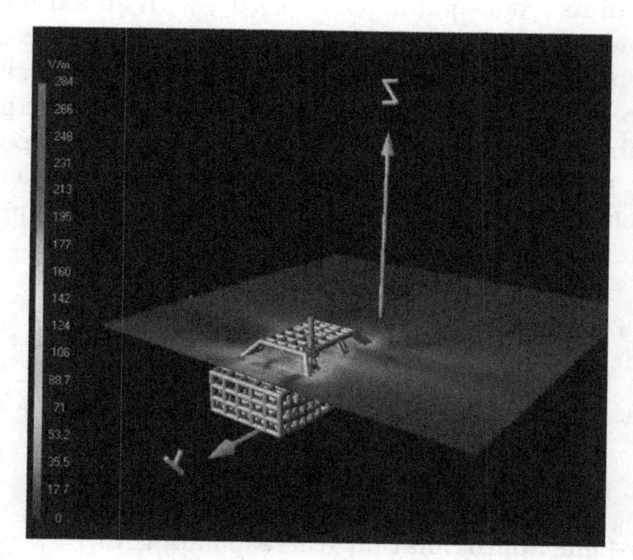


**Figure 3.17.** *Example of a digital study of a vehicle-mounted antenna in near-field. For a color version of the figure, see www.iste.co.uk/patin/power4.zip*

### 3.3.5.4. *Magnetic dipoles*

A magnetic dipole is a conducting closed loop of radius $R \ll \lambda$. In these conditions, the current $I = I_0.e^{j\omega t}$ circulating in the dipole may be presumed to be the same at all points in the loop.

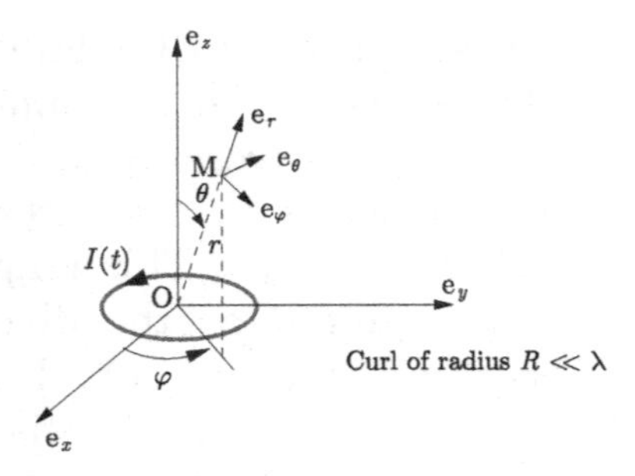


**Figure 3.18.** *Magnetic dipole aligned with the line* $(O, \mathbf{e}_z)$

Figure 3.18 shows this dipole located at the origin of a Cartestian reference plane $(O, \mathbf{e}_x, \mathbf{e}_y, \mathbf{e}_z)$ with an axis oriented along the line $(O, \mathbf{e}_z)$. The aim is to evaluate the resulting electromagnetic field at a point $M$ located at a distance $r$ from the origin. The electromagnetic field is then expressed using spherical coordinates $(\mathbf{e}_r, \mathbf{e}_\theta, \mathbf{e}_\varphi)$ linked to this point $M$. In this reference frame, the electric field has a single component with axis $\mathbf{e}_\varphi$, while the magnetic field presents components following $\mathbf{e}_r$ and $\mathbf{e}_\theta$:

$$\begin{cases} E_\varphi = -j\frac{\omega^3\mu_0\varepsilon_0 I_0 R^2}{4}\sin\theta\cdot\left(j\frac{1}{2\pi}\left(\frac{\lambda}{r}\right)+\frac{1}{(2\pi)^2}\left(\frac{\lambda}{r}\right)^2\right).e^{j\omega\left(t-\frac{r}{c}\right)} \\ H_r = j\frac{\omega^3\mu_0\varepsilon_0 I_0 R^2}{2Z_0}\cos\theta\cdot\left(\frac{1}{(2\pi)^2}\left(\frac{\lambda}{r}\right)^2-\frac{1}{(2\pi)^3}\left(\frac{\lambda}{r}\right)^3\right).e^{j\omega\left(t-\frac{r}{c}\right)} \\ H_\theta = j\frac{\omega^3\mu_0\varepsilon_0 I_0 R^2}{4Z_0}\cos\theta\cdot\left(j\frac{1}{2\pi}\left(\frac{\lambda}{r}\right)+\frac{1}{(2\pi)^2}\left(\frac{\lambda}{r}\right)^2-\frac{1}{(2\pi)^3}\left(\frac{\lambda}{r}\right)^3\right).e^{j\omega\left(t-\frac{r}{c}\right)} \end{cases}$$

[3.136]

A far-field approximation (corresponding to $r \gg \frac{\lambda}{2\pi}$) can then be established, with terms in $1/r^2$ and $1/r^3$ which become negligible, leading to the following result:

$$\begin{cases} E_\varphi = -j\frac{\omega^3\mu_0\varepsilon_0 I_0 R^2}{4}\sin\theta\cdot j\frac{1}{2\pi}\left(\frac{\lambda}{r}\right)\cdot e^{j\omega\left(t-\frac{r}{c}\right)} \\ H_r = 0 \\ H_\theta = j\frac{\omega^3\mu_0\varepsilon_0 I_0 R^2}{4Z_0}\cos\theta\cdot j\frac{1}{2\pi}\left(\frac{\lambda}{r}\right)\cdot e^{j\omega\left(t-\frac{r}{c}\right)} \end{cases} \qquad [3.137]$$

In this case, the term of the radial magnetic field disappears, and only transversal terms (i.e. perpendicular to the direction of propagation) remain: this is a TEM wave. Moreover, it is possible to verify that the ratio $E_\varphi/H_\theta$ is equal to the intrinsic impedance of a vacuum $Z_0$.

In near-field, however, higher-order terms become most important (in $1/r^2$ for $E_\varphi$ and in $1/r^3$ for $H_r$ and $H_\theta$):

$$\begin{cases} E_\varphi = -j\frac{\omega^3\mu_0\varepsilon_0 I_0 R^2}{4}\sin\theta\cdot\frac{1}{(2\pi)^2}\left(\frac{\lambda}{r}\right)^2\cdot e^{j\omega\left(t-\frac{r}{c}\right)} \\ H_r = -j\frac{\omega^3\mu_0\varepsilon_0 I_0 R^2}{2Z_0}\cos\theta\cdot\frac{1}{(2\pi)^3}\left(\frac{\lambda}{r}\right)^3\cdot e^{j\omega\left(t-\frac{r}{c}\right)} \\ H_\theta = -j\frac{\omega^3\mu_0\varepsilon_0 I_0 R^2}{4Z_0}\cos\theta\cdot\frac{1}{(2\pi)^3}\left(\frac{\lambda}{r}\right)^3\cdot e^{j\omega\left(t-\frac{r}{c}\right)} \end{cases} \quad [3.138]$$

The electromagnetic field is made up of two orthogonal components, but these components are no longer perpendicular to the direction of propagation. As in the case of the electric dipole, the configuration of the near-field is more complex than that of the far-field, and the magnetic field is dominant.

## 3.4. Natural radiated interference

Natural sources of radiation exist which have the ability to cause failures in electronic equipment. This is particularly true in space, where satellites are exposed to cosmic radiation, and more precisely to the electromagnetic waves and particles emitted by the sun. As we will see in the following section, equipment is protected by shielding, but this is problematic due to the additional weight component. Moreover, while thin shielding is sufficient for protection against electromagnetic waves encountered on earth, thicker shielding is required for protection against high-energy waves (similar to those produced by radioactive material – i.e. X and $\gamma$ rays) and particles moving at high speed.

Note, moreover, that equipment sent into space is often made up of hardened components with their own integrated protection (in the form of materials and/or software[7]).

REMARK 3.13.– The protection methods discussed here for natural radiation encountered in space are also suitable for use in applications subject to high levels of artificial radiation, such as those encountered in the nuclear sector, whether civilian (for example, for maintenance purposes in nuclear power stations) or military.

## 3.5. Protection

Protection against radiated interference generally takes the form of shielding. Shielding consists of placing a metallic screen between the interference source and the victim: the shield does not necessarily have to be very thick, but it should, ideally, be hermetic.

Hermetic casings are able to block a very wide range of frequencies – particularly visible light.

However, this raises a number of practical issues:

– casings often include openings for the passage of cables, connectors or buttons;

– certain openings are required to allow air to circulate (ventilation for cooling purposes).

Generally speaking, these openings constitute potential points for the introduction of disturbances, with a possible risk of diffraction. Note, however, that the interference requiring blocking may be located in a range of frequencies

7 Processors designed for aeronautic and aerospace applications can include error correction codes in their own machine language in order to minimize the risk of bugs, resulting from interference caused by ambient radiation.

which is compatible with perforated shielding. This is particularly interesting not only from a thermal perspective, but also from an economic perspective (less material is used, reducing costs) and in terms of equipment weight. This is seen in certain satellite dishes, such as that shown in Figure 3.19, which are perforated in order to reduce wind resistance.

**Figure 3.19.** *Perforated satellite dish*

From a theoretical perspective, considering an ideal conduction screen (with infinite conductivity), it is necessary to show that the electric and magnetic fields are zero inside the material. Mathematically (and physically) speaking, an incident wave of an electric (or magnetic) field is compensated on the surface of the screen by a wave moving in the opposite direction (such as a reflected wave). This wave is generated by the fields and currents induced by the incident field. In fact, the electric field is reflected by a network of fine, parallel conducting wires (with a low separation distance $e$ in relation to the wavelength $\lambda$), co-linear with the direction of E. This is known as a wire-grid polarizer. If two perpendicular grids are assembled to produce a conducting mesh, this mesh constitutes an electromagnetic screen which is effective for waves with a length greater than the mesh (characteristic dimension $e$). For instance, considering a mesh with sides 1 mm, if the screen is effective for waves of a length ten times higher (e.g. 1 cm), the screen may be considered suitable to reflect waves with frequencies less than or equal to 30 GHz. More generally, when there is an opening in a conducting screen, the electromagnetic field will not be able to pass effectively through the gap if the perimeter of the opening is small in relation to $\lambda$. However, if the two parameters are of the same order of magnitude, the opening may operate as an antenna (i.e. a radiating opening) and the shield will be completely ineffective (or even constitute a reverberating box).

REMARK 3.14.– The shielding must be thicker than the skin thickness (see equation [5.47], in Chapter 5, Volume 1 [PAT 15a]). Shielding is consequently simpler (lighter and less costly) for high frequencies than for low frequencies. In the low-frequency domain (particularly in electrical engineering and power electronics), the magnetic field is most problematic: for this reason, shielding materials with high magnetic permeability are selected, particularly mu-metal (essentially a nickel, iron and molybdenum alloy) which is designed for this type of application. The aim is to direct the field into the shielding wall to prevent it from entering the protected zone; for this to succeed, the material must not reach saturation.

## 3.6. Experimental aspects

Experiments concerning radiated EMC require the use of specific equipment. The radiation generated by a device must be evaluated using antenna (of variable type according to the frequency range under study); for analysis to succeed, the equipment must be placed in an anechoic chamber, itself generally situated inside a shielded room in order to avoid not only the reflection of emitted waves by the walls, but also radiation from the outside environment.

Figure 3.20 shows an electromagnetic anechoic chamber, padded with pyramids of polyurethane foam, loaded with carbon in order to absorb waves. These types of chamber may also use ferrite tiles for the same purposes (often at different frequencies and power levels).

Figure 3.21 shows a shielded room, which operates as a Faraday cage over a wide range of frequencies. The walls, floor and ceiling of a room of this type are covered in conducting sheets (generally copper), forming an almost perfect equipotential.

This kind of environment may be considered to be perfectly insulated from the external electric field, and, in the absence of a source, the field within the room is null[8]. To avoid disturbing the electromagnetic field within an installation of this type, 100% wooden furniture is preferred.

Finally, for sensitivity tests, we do not necessarily aim to characterize a preferred direction in which a component will or will not be sensitive to an electromagnetic field (except in the case of isolated equipment made up of polluting and/or sensitive elements), but rather to verify whether a device or system as a whole is resistant to a given disturbance. To carry out tests of this type in an anechoic chamber by moving an emitter antenna around the object would be costly and labor-intensive; it is therefore better to use an over-dimensioned mode-mixing echo chamber. This allows us to obtain multiple resonances in given directions, in order to apply isotrope solicitations to the "victim". The metallic walls in this type of chamber reflect electromagnetic waves, and a mode shuffler, made from a motorized pole supporting reflecting poles, is used; this device is shown in the in

8 This is not the case in open air, as the electric field at ground level in clear weather is typically of 100 V/m, a value which increases considerably in the vicinity of angular conductors (peak effect).

Figure 3.22. Immunity (or sensitivity) standards apply to this type of test installation, notably EN61000-4-21.

**Figure 3.20.** *Electromagnetic anechoic chamber*

**Figure 3.21.** *Shielded room*

**Figure 3.22.** *Mode-shuffling echo chamber*

## 3.7. Practical applications

Readers may wonder how these wave propagation tools (in free space or using waveguides) are useful in the context of power electronics. It is not easy to use all of these theoretical and practical tools, designed for radio frequency applications (and mostly used in information transmission, wirelessly or using coaxial cables, or microribbon lines on PCBs). Note, however, that the use of these tools becomes apparent if we ask the right questions concerning electric diagram models of equipment, for which EMC filters are required, or where electromagnetic interference is to be analyzed.

If these aspects are not taken into consideration, there is a risk of falling into the usual trap of considering electronics as a type of black magic (not related to the sub-title of the Smith chart presented in Figure 3.4). The preliminary questions to be asked before embarking on a study, and particularly before modeling is attempted, may be summarized as follows:

– up to what frequency $f_{\max}$ should the system be analyzed?

– what are the supports for the propagation of electrical quantities (currents and voltages) – more precisely, is there a dielectric with a permittivity $\varepsilon = \varepsilon_0.\varepsilon_r$ which is significantly different than that of a vacuum (i.e. $\varepsilon_r \neq 1$)?

– based on the first two responses, what is the minimum wavelength $\lambda_{\min} = \frac{v}{f_{\max}}$ (where $v = \frac{c}{\sqrt{\varepsilon_r}}$[9]) associated with the transported voltages/currents?

– if the lengths involved in the system are greater than 10% of the wavelength[10], phenomena may appear which are difficult to explain using distributed constant models; consequently, predictions made using a model of this type will be imprecise and subject to errors.

In practice, the aim is to minimize the length over which disturbances are propagated, notably in the case of common-mode interference, which can easily overcome shielding. A circulation loop with a common-mode current constitutes an excellent antenna, both in terms of radiation and of sensitivity to external pollution. It is therefore generally advisable to avoid distributed constant phenomena in power electronics. Note, however, that applications connected to the network are increasingly confronted with this type of phenomena, due to the widespread use of the electrical network for information transmission (service signals sent by the French electricity network operator, RTE, for example), but also, and especially, due to Ethernet modules with carrier currents. In the context of these applications, "propagation" aspects are essential in understanding and implementing this type of equipment, which needs to cohabit successfully with the electronic power converters present in any modern domestic or industrial setting.

9 Note that $c$ is the speed of light in a vacuum, i.e. approximately $3 \times 10^8$ m/s.

10 This is a rule of thumb with a somewhat blurred boundary. Certain more conservative authors consider that this boundary lies closer to $\lambda_{\min}/20$, for example.

# Appendix 1

# Formulas for Electrical Engineering and Electromagnetism

## A1.1. Sinusoidal quantities

### A1.1.1. *Scalar signals*

#### A1.1.1.1. *Definitions*

Sinusoidal waveforms are extremely widespread in electrical engineering, both for voltages and for currents. In this case, we will consider a generic signal of the form:

$$x(t) = X_{\max} \cos(\omega t - \varphi) \quad \text{[A1.1]}$$

This real signal is associated with an equivalent complex signal:

$$\underline{x}(t) = X_{\max} . e^{j(\omega t - \varphi)} \quad \text{[A1.2]}$$

This vector may be represented in the complex plane. We obtain a circular trajectory of radius $X_{\max}$ with a vector rotating at a constant speed $\omega$ in a counterclockwise direction. This representation (which is widespread in electrical engineering) is known as a Fresnel diagram (or, more simply, a vector diagram).

REMARK A1.1.– Derivation and integration calculations are greatly simplified in the complex plane, as they are replaced, respectively, by multiplying or dividing by $j\omega$. To return to the real domain, we simply take the real part of the corresponding complex signal: $x(t) = \mathfrak{Re}\left[\underline{x}\,(t)\right]$.

The rotating component $e^{j\omega t}$ of the complex vectors is meaningless when studying linear circuits; the *amplitudes* and the relative phases between the different quantities under study are the only important elements. Note that an absolute phase for a sinusoidal value would be meaningless; the choice of a reference value of the form $X_{\text{ref}}.\cos(\omega.t)$, associated with the vector $X_{\text{ref}}.e^{j\omega t}$, is purely arbitrary.

Complex vectors are also often represented (in the literature) using the RMS value of the real value in question as the modulus, and not the real amplitude.

A1.1.1.2. *Trigonometric formulas*

When making calculations using complex values, we need Euler's formulas:

$$\begin{cases} \cos\theta = \frac{e^{j\theta}+e^{-j\theta}}{2} \\ \sin\theta = \frac{e^{j\theta}-e^{-j\theta}}{2j} \end{cases} \quad [A1.3]$$

These two formulas can be used to give the four basic trigonometric formulas used in electrical engineering:

$$\begin{cases} \cos(a+b) = \cos a\cos b - \sin a\sin b \\ \cos(a-b) = \cos a\cos b + \sin a\sin b \\ \sin(a+b) = \sin a\cos b + \cos a\sin b \\ \sin(a-b) = \sin a\cos b - \cos a\sin b \end{cases} \quad [A1.4]$$

These four equations allow us to establish four further equations:

$$\cos a\cos b = \frac{1}{2}\left(\cos(a+b) + \cos(a-b)\right) \quad [A1.5]$$

$$\begin{cases} \cos a \cos b = \frac{1}{2}\left(\cos(a+b) + \cos(a-b)\right) \\ \sin a \sin b = \frac{1}{2}\left(\cos(a-b) - \cos(a-b)\right) \\ \sin a \cos b = \frac{1}{2}\left(\sin(a+b) + \sin(a-b)\right) \\ \cos a \sin b = \frac{1}{2}\left(\sin(a+b) - \sin(a-b)\right) \end{cases} \quad [A1.6]$$

## A1.1.2. *Vector signals (three-phase context)*

A1.1.2.1. *Reference frame* $(a, b, c)$

Three-phase systems are very much common in electrical engineering, particularly balanced three-phase systems. A vector $(\mathbf{x}_3) = (x_a, x_b, x_c)^t$ with three balanced components is therefore expressed as:

$$(\mathbf{x}_3) = X_{\max} \begin{pmatrix} \cos\theta \\ \cos\left(\theta - \frac{2\pi}{3}\right) \\ \cos\left(\theta + \frac{2\pi}{3}\right) \end{pmatrix} \text{ where } \theta = \omega.t + \phi_0 \quad [A1.7]$$

in the case of a direct system, or:

$$(\mathbf{x}_3) = X_{\max} \begin{pmatrix} \cos\theta \\ \cos\left(\theta + \frac{2\pi}{3}\right) \\ \cos\left(\theta - \frac{2\pi}{3}\right) \end{pmatrix} \text{ where } \theta = \omega.t + \phi_0 \quad [A1.8]$$

in the inverse case.

DEFINITION A1.1.– A balanced three-phase system is thus made up of three sinusoids of the same amplitude and same frequency, with a phase deviation of $\frac{2\pi}{3}$.

A direct three-phase system is characterized by the fact that, taking phase 1 as a reference point (i.e. first component), the second component has a delay of 120° (in a balanced situation) and the third component presents a delay of 120° in relation to the second component.

An inverse three-phase system is characterized by the fact that, taking phase 1 as a reference point (i.e. first component), the third component has a delay of 120° (in a balanced situation) and the second component presents a delay of 120° in relation to the third component. A direct system can be converted into an inverse system (and vice versa) by permutations of two components.

A1.1.2.2. *Three-phase to two-phase transformation* $(\alpha, \beta)$

It is important to note that a balanced three-phase system (whether direct or inverse) presents an important property in that the sum of the components is null:

$$x_a + x_b + x_c = 0 \quad \text{[A1.9]}$$

This sum is classically referred to as the zero sequence component (denoted as $x_0$). A balanced three-phase system is therefore not linearly independent in that, given two of the components, we may calculate the value of the third component. It is therefore possible to propose a three-phase to two-phase transformation without any information loss. The simplest transformation, known as the Clarke (abc-to-$\alpha\beta$) transformation, allows us to associate an initial vector $(\mathbf{x}_3) = (x_a, x_b, x_c)^t$ with an equivalent two-phase vector $(\mathbf{x}_{\alpha\beta}) = (\mathbf{x}_2) = (x_\alpha, x_\beta)^t$ using components of the same amplitude as those in the initial vector. This operation introduces the Clarke matrix $C_{32}$:

$$X_{\max} \begin{pmatrix} \cos\theta \\ \cos\left(\theta + \frac{2\pi}{3}\right) \\ \cos\left(\theta - \frac{2\pi}{3}\right) \end{pmatrix} = X_{\max} \cdot \underbrace{\begin{pmatrix} 1 & 0 \\ -1/2 & \sqrt{3}/2 \\ -1/2 & -\sqrt{3}/2 \end{pmatrix}}_{C_{32}} \cdot \begin{pmatrix} \cos\theta \\ \sin\theta \end{pmatrix} \quad \text{[A1.10]}$$

This gives the following direct transformation:

$$(\mathbf{x}_3) \triangleq C_{32} \cdot (\mathbf{x}_2) \quad \text{[A1.11]}$$

The Clarke transformation may be extended by taking account of the zero sequence component $x_0$, presented in [A1.9]:

$$(\mathbf{x}_3) \triangleq C_{32}.(\mathbf{x}_2) + C_{31}.x_0 \qquad \text{[A1.12]}$$

with:

$$C_{31} = \begin{pmatrix} 1 \\ 1 \\ 1 \end{pmatrix} \qquad \text{[A1.13]}$$

Noting certain properties of matrices $C_{32}$ and $C_{31}$:

$$\begin{array}{l} C_{32}^t C_{32} = \frac{3}{2}\begin{pmatrix} 1\,0 \\ 0\,1 \end{pmatrix} \;;\; C_{31}^t C_{31} = 3 \\ C_{32}^t C_{31} = \begin{pmatrix} 0 \\ 0 \end{pmatrix} \;;\; C_{31}^t C_{32} = (0\,0) \end{array} \qquad \text{[A1.14]}$$

we can establish the inverse transformation:

$$(\mathbf{x}_2) \triangleq \frac{2}{3} C_{32}^t.(\mathbf{x}_3) \qquad \text{[A1.15]}$$

and:

$$x_0 \triangleq \frac{1}{3} C_{31}^t.(\mathbf{x}_3) \qquad \text{[A1.16]}$$

A1.1.2.3. *Concordia variant*

A second three-phase to two-phase transformation is also widely used in the literature, with properties similar to those of the Clarke transformation. This variation does not retain the amplitudes of the transformed values, but allows us to retain powers. This operation is known as the Concordia transformation and is based on two matrices, denoted $T_{32}$ and $T_{31}$, deduced from $C_{32}$ and $C_{31}$:

$$T_{32} = \sqrt{\frac{2}{3}} C_{32} \;;\; T_{31} = \frac{1}{\sqrt{3}} C_{31} \qquad \text{[A1.17]}$$

The properties of these matrices are deduced from those established in [A1.14]:

$$\begin{aligned} T_{32}^t T_{32} &= \begin{pmatrix} 1 & 0 \\ 0 & 1 \end{pmatrix} \; ; \; T_{31}^t T_{31} = 1 \\ T_{32}^t T_{31} &= \begin{pmatrix} 0 \\ 0 \end{pmatrix} \; ; \; T_{31}^t T_{32} = \begin{pmatrix} 0 & 0 \end{pmatrix} \end{aligned} \qquad \text{[A1.18]}$$

This produces a direct transformation of the form:

$$(\mathbf{x}_3) \triangleq T_{32}.\,(\mathbf{x}_2) + T_{31}.x_0 \qquad \text{[A1.19]}$$

with the following inverse transformation:

$$(\mathbf{x}_2) \triangleq T_{32}^t.\,(\mathbf{x}_3) \qquad \text{[A1.20]}$$

and:

$$x_0 \triangleq T_{31}^t.\,(\mathbf{x}_3) \qquad \text{[A1.21]}$$

A1.1.2.4. *Park transformation*

The Park (abc-to-dq) transformation consists of associating the Clarke (or Concordia) transformation with a rotation in the two-phase reference plane $(\alpha, \beta)$ onto a reference frame $(d, q)$. This operation is carried out using the rotation matrix $P(\theta)$, defined as:

$$P(\theta) = \begin{pmatrix} \cos\theta & -\sin\theta \\ \sin\theta & \cos\theta \end{pmatrix} \qquad \text{[A1.22]}$$

Thus, if we associate a vector $(\mathbf{x}_{dq}) = (x_d, x_q)^t$ with the initial two-phase vector $(\mathbf{x}_{\alpha\beta}) = (\mathbf{x}_2)$ (obtained from a Clarke or Concordia transformation), we obtain the following relationship:

$$(\mathbf{x}_{\alpha\beta}) = (\mathbf{x}_2) \triangleq P(\theta).\,(\mathbf{x}_{dq}) \qquad \text{[A1.23]}$$

The choice of a frame of reference involves the definition of angle $\theta$, selected arbitrarily. Generally, the chosen reference frame is synchronous with the rotating values (sinusoidal components with an angular frequency $\omega$), but this is not obligatory.

The following (non-exhaustive) list shows a number of properties of matrix $P(\theta)$:

$$P(0) = \begin{pmatrix} 1 & 0 \\ 0 & 1 \end{pmatrix} = \mathbb{I}_2 \; ; \; P\left(\tfrac{\pi}{2}\right) = \begin{pmatrix} 0 & -1 \\ 1 & 0 \end{pmatrix} = \mathbb{J}_2 \text{ such that } \mathbb{J}_2 = -\mathbb{I}_2 \qquad \text{[A1.24]}$$

$$P(\alpha + \beta) = P(\beta + \alpha) = P(\alpha) . P(\beta) = P(\beta) . P(\alpha) \qquad \text{[A1.25]}$$

$$P(\alpha)^{-1} = P(\alpha)^t = P(-\alpha) \qquad \text{[A1.26]}$$

$$\tfrac{d}{dt}\left[P(\alpha)\right] = \tfrac{d\alpha}{dt} \cdot P\left(\alpha + \tfrac{\pi}{2}\right) = \tfrac{d\alpha}{dt} \cdot P(\alpha) \cdot P\left(\tfrac{\pi}{2}\right) = \mathbb{J}_2 \tfrac{d\alpha}{dt} \cdot P(\alpha) \qquad \text{[A1.27]}$$

A1.1.2.5. *Phasers or complex vectors*

The matrix formalism of the Clarke, Concordia and Park transformations may be replaced by an equivalent complex representation. Evidently, a rotation of the frame of reference by angle $\theta$ may be obtained by using a complex coefficient $e^{j\theta}$ as easily as with a rotation matrix $P(\theta)$. To this end, we use a "phaser" $\underline{x}_s$ defined in a stationary frame of reference:

$$\underline{x}_s = x_\alpha + j.x_\beta \qquad \text{[A1.28]}$$

The phaser is also defined in a rotating frame ($\underline{x}_r$):

$$\underline{x}_r = x_d + j.x_q \qquad \text{[A1.29]}$$

Note that these complex representations may be obtained using matrix transformations. The real transformations seen in the previous sections each have an equivalent complex transformation, as shown in Table A1.1.

| Real transformation | Complex transformation |
|---|---|
| Clarke | Fortescue |
| Concordia | Lyon |
| Park | Ku |

**Table A1.1.** *Correspondence between real and complex transformations (names)*

## A1.2. General characteristics of signals in electrical engineering

This section presents the formulas used for calculating the *general characteristics of periodic signals* traditionally encountered in electrical engineering. However, it does not cover formulas related to spectral analysis, which are covered in Appendix 2 in this volume and in Volume 2 [PAT 15b].

In this section, we will therefore cover the formulas used to calculate the average and RMS values of given quantities, applied to two widespread signal types: sinusoids and the asymmetric square signal of duty ratio $\alpha$.

### A1.2.1. *Average value*

#### A1.2.1.1. *General definition*

The average value $\langle x \rangle$ of a $T$-periodic signal $x(t)$ is defined generally by the integral:

$$\langle x \rangle = \frac{1}{T} \int_0^T x(t).dt \qquad \text{[A1.30]}$$

REMARK A1.2.– In this case, the integration limits are chosen arbitrarily. Only the interval between the two limits is important, and it must be equal to $T$.

#### A1.2.1.2. *Sinusoids*

In the case of sinusoids, we evidently obtain an average value of zero.

A1.2.1.3. *Asymmetric square*

The $T$-periodic asymmetric square $x(t)$ studied here has a certain value $X_0$ during a period $\alpha T$, then 0 for the rest of the period. We can therefore write the average value $\langle x \rangle$ directly:

$$\langle x \rangle = \frac{1}{T}\int_0^T x(t).dt = \frac{1}{T}\int_0^{\alpha T} X_0.dt = \alpha.X_0 \qquad \text{[A1.31]}$$

### A1.2.2. *RMS value*

A1.2.2.1. *General definition*

The RMS value $X_{\text{rms}}$ of a $T$-periodic signal $x(t)$ is defined generally by the integral:

$$X_{\text{rms}} = \sqrt{\frac{1}{T}\int_0^T x^2(t).dt} \qquad \text{[A1.32]}$$

REMARK A1.3.– When calculating the average value, the integration limits are chosen arbitrarily. Only the interval between the two limits is important, and it must be equal to $T$.

A1.2.2.2. *Sinusoids*

For a sinusoid of amplitude $X_{\text{max}}$, the RMS value is $X_{\text{rms}} = \frac{X_{\text{max}}}{\sqrt{2}}$.

A1.2.2.3. *Asymmetric square*

The $T$-periodic asymmetric square $x(t)$ defined in section A1.2.1 presents an RMS value expressed as:

$$X_{\text{rms}} = \sqrt{\frac{1}{T}\int_0^{\alpha T} X_0^2.dt} = \sqrt{\alpha}.X_0 \qquad \text{[A1.33]}$$

## A1.3. Energy and power

### A1.3.1. *Energy*

In mechanics, energy is obtained by the operation of a force over a certain distance. In electrical engineering, this term corresponds to the movement of a charge following a variation in electrical potential. In particle physics, a unit known as an electron-volt (eV) is used for energy values at the atomic level. The energy formulas used in power electronics (expressed in Joules (J)) correspond to the energy stored in an inductance or a capacitor.

In an inductance, the energy $E_L$ (magnetic energy) depends on the current $I$ and the inductance $L$:

$$E_L = \frac{1}{2}LI^2 \qquad [A1.34]$$

For a capacitor, the energy $E_C$ (electrostatic energy) depends on the voltage $V$ and the capacitance $C$:

$$E_C = \frac{1}{2}CV^2 \qquad [A1.35]$$

### A1.3.2. *Instantaneous power*

The instantaneous power $p(t)$ given – or provided to – the dipole is linked, according to the passive sign convention (PSC), to the voltage $v(t)$ at its terminals and the current $i(t)$ passing through it as follows:

$$p(t) = v(t).i(t) \qquad [A1.36]$$

This power is defined in watts (W). It is linked to the energy consumed $E$ (in J) between two instants $t_1$ and $t_2$ by the following integral:

$$E = \int_{t_1}^{t_2} p(t).dt \qquad [A1.37]$$

The instantaneous power $p(t)$ is connected to the variation in energy $e(t)$ which can also be defined (up to an additive constant) as a function of time. In this case, we obtain:

$$p(t) = \frac{de(t)}{dt} \qquad [A1.38]$$

## A1.3.3. *Average power*

As for any $T$-periodic signal, the average power $P$ is obtained using the following formula:

$$P = \frac{1}{T}\int_0^T p(t).dt = \frac{1}{T}\int_0^T v(t).i(t).dt \qquad [A1.39]$$

In the case of a resistive charge $R$, we can establish the following relationship (Ohm's law):

$$v(t) = R.i(t) \qquad [A1.40]$$

This allows us to formulate two possible expressions for this power:

$$P = \frac{R}{T}\int_0^T i^2(t).dt = R.I_{\text{rms}}^2 \qquad [A1.41]$$

and:

$$P = \frac{1}{RT}\int_0^T v^2(t).dt = \frac{V_{\text{rms}}^2}{R} \qquad [A1.42]$$

where $V_{\text{rms}}$ and $I_{\text{rms}}$ are the RMS values of the voltage and the current, respectively.

## A1.3.4. *Sinusoidal mode*

### A1.3.4.1. *Single phase*

In single phase sinusoidal operating mode, we can, generally speaking, consider a voltage $v(t)$ of the form:

$$v(t) = V_{\text{rms}}\sqrt{2}\cos(\omega t) \quad \text{[A1.43]}$$

as the phase reference, with a current, with a phase deviation angle $\varphi$ (the lag in relation to the voltage), expressed as:

$$i(t) = I_{\text{rms}}\sqrt{2}\cos(\omega t - \varphi) \quad \text{[A1.44]}$$

Calculating the instantaneous power obtained using these two values, we obtain:

$$p(t) = V_{\text{rms}}I_{\text{rms}}\left(\cos(2\omega t - \varphi) + \cos\varphi\right) \quad \text{[A1.45]}$$

We thus obtain two terms:

– a constant term, which is, evidently, the average power, referred to in this context as active power;

– a variable term, with an angular frequency of $2\omega$, known as fluctuating power.

The first interesting result is, therefore, the expression of the average (active) power $P$:

$$P = V_{\text{rms}}I_{\text{rms}}\cos\varphi \quad \text{[A1.46]}$$

In terms of voltage dimensioning (thickness of insulation) and current dimensioning (cross-section of conductors) of equipment, the real power value used for design purposes is known as the apparent power $S$ , and is obtained by directly multiplying the RMS voltage value by the RMS current value:

$$S = V_{\text{rms}}I_{\text{rms}} \quad \text{[A1.47]}$$

To emphasize the "fictional" character of this power, it is not given in W, but in volt-amperes (VA).

In electrical engineering, we then use the notion of *reactive power* $Q$, which allows us to establish a connection between the active power $P$ and the apparent power $S$. This is expressed as:

$$Q = V_{\text{rms}} I_{\text{rms}} \sin \varphi \qquad [A1.48]$$

The connection between $P$, $Q$ and $S$ is thus:

$$S^2 = P^2 + Q^2 \qquad [A1.49]$$

As in the case of apparent power, this power value is fictional; it is not measured in W, or in VA, but rather in volt ampere reactive (VAR).

REMARK A1.4.– Equation [A1.49] is only valid if the voltage *and* the current are sinusoidal. In non-sinusoidal mode, we introduce an additional power, denoted $D$, known as the deformed power. This is used to establish a new equation as follows:

$$S^2 = P^2 + Q^2 + D^2 \qquad [A1.50]$$

The instantaneous power is always positive (respectively, negative) when $\varphi = 0°$ (respectively, $\varphi = 180°$), but if $\varphi$ takes a different value, $p(t)$ cancels out, changing the sign. In these conditions, the direction of transfer of electronic energy between the source and the load is reversed.

A1.3.4.2. *Three phase*

In a three-phase context, using the "voltage" vector ($\mathbf{v}_3$) as a point of reference, and more specifically as the first component, we take (based on the hypothesis of a direct balanced system):

$$(\mathbf{v}_3) = V_{\text{rms}}\sqrt{2}\begin{pmatrix}\cos(\omega t)\\ \cos\left(\omega t - \frac{2\pi}{3}\right)\\ \cos\left(\omega t + \frac{2\pi}{3}\right)\end{pmatrix} \quad \text{[A1.51]}$$

From this, we deduce the "current" vector $(\mathbf{i}_3)$, with a lag in each component when compared to the corresponding components in $(\mathbf{v}_3)$:

$$(\mathbf{i}_3) = I_{\text{rms}}\sqrt{2}\begin{pmatrix}\cos(\omega t - \varphi)\\ \cos\left(\omega t - \frac{2\pi}{3} - \varphi\right)\\ \cos\left(\omega t + \frac{2\pi}{3} - \varphi\right)\end{pmatrix} \quad \text{[A1.52]}$$

A matrix formalism may be used to obtain the expression of the instantaneous power $p(t)$:

$$p(t) = (\mathbf{v}_3)^t \,.\, (\mathbf{i}_3) \quad \text{[A1.53]}$$

In this case, the Park factorization of the "voltage and current" vectors is particularly effective:

$$(\mathbf{v}_3) = V_{\text{rms}}\sqrt{2}.C_{32}\begin{pmatrix}\cos(\omega t)\\ \sin(\omega t)\end{pmatrix}$$

$$= V_{\text{rms}}\sqrt{2}.C_{32}.P(\omega t)\,.\begin{pmatrix}1\\ 0\end{pmatrix} \quad \text{[A1.54]}$$

$$(\mathbf{i}_3) = I_{\text{rms}}\sqrt{2}.C_{32}\begin{pmatrix}\cos(\omega t - \varphi)\\ \sin(\omega t - \varphi)\end{pmatrix}$$

$$= I_{\text{rms}}\sqrt{2}.C_{32}.P(\omega t - \varphi)\,.\begin{pmatrix}1\\ 0\end{pmatrix} \quad \text{[A1.55]}$$

Hence:

$$p(t) = 2V_{\text{rms}}.I_{\text{rms}} \begin{pmatrix} 1 & 0 \end{pmatrix} .P(-\omega t).C_{32}^{t}.C_{32}.P(\omega t - \varphi).\begin{pmatrix} 1 \\ 0 \end{pmatrix} \quad \text{[A1.56]}$$

After simplification, this gives us:

$$p(t) = 3V_{\text{rms}}.I_{\text{rms}} \cos\varphi \quad \text{[A1.57]}$$

Note that we obtain the instantaneous power, and not an average value. This highlights a notable property of three-phase systems: there is no globally fluctuating power in this configuration.

The active power $P$ is therefore defined as follows:

$$P = p(t) = 3V_{\text{rms}}.I_{\text{rms}} \cos\varphi \quad \text{[A1.58]}$$

The notions of reactive power $Q$ and apparent power $S$ are also used in three-phase contexts, with the following expressions:

$$\begin{cases} Q = 3V_{\text{rms}}.I_{\text{rms}} \sin\varphi \\ S = 3V_{\text{rms}}.I_{\text{rms}} \end{cases} \quad \text{[A1.59]}$$

Relationship [A1.49] is therefore still valid in a three-phase context:

$$S^2 = P^2 + Q^2 \quad \text{[A1.60]}$$

Note that variants exist, notably where the notion of line-to-line voltage is used. Voltage $V_{\text{rms}}$ is the RMS *line-to-neutral voltage* (i.e. between the phase and the neutral); the neutral is not always accessible, so the notion of line-to-line voltage is often preferred , with an RMS voltage, denoted $U_{\text{rms}}$. In the case of a balanced three-phase system, the relationship between the RMS line-to-neutral and line-to-line voltage is:

$$U_{\text{rms}} = \sqrt{3}V_{\text{rms}} \qquad \text{[A1.61]}$$

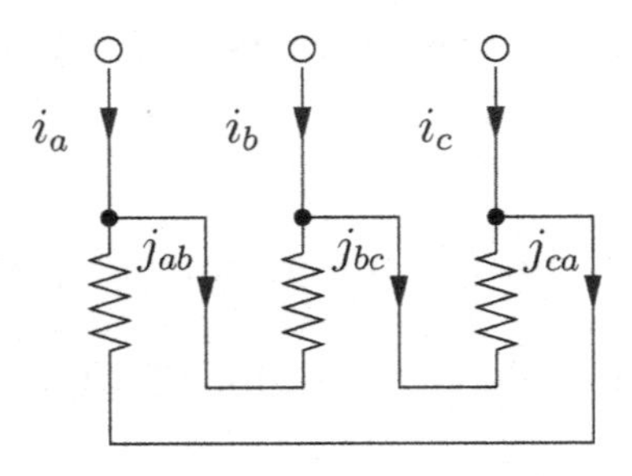


**Figure A1.1.** *Line and branch currents for a triangular connection*

A second point, which may lead to a different formulation of expression [A1.58], is concerned with currents. Generally speaking, we always have access to *line currents*, and thus to the RMS value $I_{\text{rms}}$. A second type of current can appear when using a load with a triangle connection (see Figure A1.1): this branch current presents an RMS value $J_{\text{rms}}$ with the following expression as a function of $I_{\text{rms}}$:

$$J_{\text{rms}} = \frac{I_{\text{rms}}}{\sqrt{3}} \qquad \text{[A1.62]}$$

## A1.4. Mathematics for electromagnetism

### A1.4.1. *The Green–Ostrogradsky theorem*

The Green–Ostrogradsky theorem (also known as the flux–divergence theorem) establishes a connection between the integral of the divergence of a field with vector **E** in a volume $\Omega$ and the integral of the flux of **E** on the closed surface $\partial\Omega$ delimiting the volume $\Omega$:

$$\iiint_{\Omega} \text{div}\mathbf{E}.d\omega = \oiint_{\partial\Omega} \mathbf{E} \cdot ds \qquad \text{[A1.63]}$$

where $d\omega$ is a volume element, while $ds$ is a normal vector[1] with a surface element (infinitesimal) $ds$ of the complete surface $\partial\Omega$.

### A1.4.2. *Stokes–Ampère theorem*

The Stokes–Ampère theorem establishes a connection between the the flux curl of the magnetic field **H** on a surface $\Sigma$ and the integral of the circulation of **H** along the closed contour $\partial\Sigma$ delimiting surface $\Sigma$:

$$\iint_{\Sigma} \textbf{curl}\,\mathbf{H} \cdot ds = \oint_{\partial\Sigma} \mathbf{H} \cdot dl \qquad \text{[A1.64]}$$

where $ds$ is a normal vector[2] with a surface element (infinitesimal) $ds$ of the complete surface $\Sigma$. Element $dl$ is a vector (whose norm is $dl$) tangent to the closed contour $\partial\Sigma$.

### A1.4.3. *Differential and referential operators*

The definition of the differential operators used in electromagnetism (primarily grad, div and curl) is dependent on the chosen frame of reference. Using the Cartesian

---

1 Oriented toward the outside of volume $\Omega$.

2 Oriented in accordance with the right-hand rule as a function of the choice of orientation of contour $\partial\Sigma$.

coordinate system, the nabla operator (vector), $\nabla$, allows us to easily write these operators as:

$$\nabla = \begin{pmatrix} \frac{\partial}{\partial x} \\ \frac{\partial}{\partial y} \\ \frac{\partial}{\partial z} \end{pmatrix} \quad [A1.65]$$

and we know that:

$$\begin{cases} \mathbf{grad}\, V = \nabla V \\ \operatorname{div} \mathbf{E} = \nabla \cdot \mathbf{E} \\ \mathbf{curl}\, \mathbf{H} = \nabla \times \mathbf{H} \end{cases} \quad [A1.66]$$

where the symbol "·" is the scalar product and "×" is the vector product.

If we want to write these operators using spherical or cylindrical coordinates, the $\nabla$ operator is no longer suitable; in these cases, it is better to use intrinsic definitions (which are independent of the chosen frame of reference). For the gradient, we have:

$$dV = (\mathbf{grad}\, V) \cdot d\mathbf{r} \quad [A1.67]$$

where $dV$ is the exact total differential of $V$ and dr is an infinitesimal shift (vector) away from the considered point in the space (defined by vector r from the origin of the reference frame).

For the "divergence" and "curl" operators, we simply use the two theorems presented in sections A1.4.1 and A1.4.2. First, we obtain:

$$d\phi = \operatorname{div} \mathbf{E}.d\omega \quad [A1.68]$$

where $d\phi$ is the flux of **E** across the surface of the volume $d\omega$ under consideration.

We can then write:

$$dC = \mathbf{curl\,H} \cdot \mathrm{n}.dS \qquad [A1.69]$$

where $dC$ is the circulation of field H along a closed contour enclosing a surface $dS$, and with an orientation allowing us to define a normal (unitary) vector n (in accordance with the right-hand rule).

# Appendix 2

# Elements of Spectral Analysis

## A2.1. Periodic signals

### A2.1.1. *Fourier series decomposition*

A Fourier series decomposition consists of writing a $T$-periodic signal $x(t)$ (i.e. with a frequency $F = 1/T$) as an infinite (discrete) sum of sinusoids with frequency $k.F$ (where $k \in \mathbb{N}$). This Fourier series decomposition is convergent at all points, on the condition that certain mathematical conditions are met; in this context, we will consider these conditions to be met by ensuring the continuity of signal $x(t)$. Mathematically speaking, in the opposite case, convergence is not guaranteed but is still "almost always" obtained[1]. Thus, we may use the following equation:

$$x(t) = a_0 + \sum_{k=1}^{+\infty} a_k . \cos(2\pi kF.t) + b_k . \sin(2\pi kF.t) \qquad [A2.1]$$

1 While this definition is simplistic from a mathematical perspective, it is largely sufficient when studying power electronics, and may be used more widely in electrical engineering.

where (taking $k > 1$):

$$a_0 = \frac{1}{T}\int_0^T x(t).dt \tag{A2.2}$$

$$a_k = \frac{2}{T}\int_0^T x(t).\cos(2\pi kFt).dt \tag{A2.3}$$

$$b_k = \frac{2}{T}\int_0^T x(t).\sin(2\pi kFt).dt \tag{A2.4}$$

Note that the amplitude $A_k$ of the sinusoid of frequency $k.F$ is obtained by combining the "cos" and "sin" terms, i.e.:

$$A_k = \sqrt{a_k^2 + b_k^2} \tag{A2.5}$$

Another Fourier series formulation is possible, using the complex exponential $e^{j2k\pi Ft}$, where $k \in \mathbb{Z}$, in the place of the "cos" and "sin" functions:

$$x(t) = \sum_{k=-\infty}^{+\infty} c_k.e^{j2k\pi Ft} \tag{A2.6}$$

with:

$$c_k = \frac{1}{T}\int_0^T x(t).e^{j2k\pi Ft}.dt \tag{A2.7}$$

Note that in this case the $c_k$ coefficients of the Fourier series are complex numbers.

### A2.1.2. *Properties*

#### A2.1.2.1. *Symmetries*

In the case of an even $x(t)$ signal, i.e. such that $x(-t) = x(t)$, it is easy to verify that:

$$\forall k \in \mathbb{N},\ b_k = 0 \tag{A2.8}$$

For an odd $x(t)$ signal, such that $x(-t) = -x(t)$, it is easy to verify that:

$$\forall k \in \mathbb{N},\ a_k = 0 \qquad \text{[A2.9]}$$

More specifically, if we have symmetry "in $T/2$ and $T/4$", the Fourier series decomposition is simplified, with a reduction in the number of coefficients to calculate.

If (axial) symmetry exists in $T/2$, then the decomposition only includes non-null odd coefficients ($a_{2p} = b_{2p} = 0$).

In the same way, if (central) symmetry exists in $T/4$, then the odd coefficients which are multiples of 3 will be null (coefficients 3, 9, 15, etc.).

For example, consider the case of the waveform of the line-to-line voltage output of a three-phase inverter under "full wave" command (see Chapter 2, Figure 2.13, of Volume 2 [PAT 15b]). In this case, both types of symmetry are present, and only the odd components which are not multiples of 3 will have non-null amplitudes. This is easy to verify for learning purposes.

More generally, we may wish to consider the properties of the complex Fourier series decomposition. In this case, we note that the ranks are relative integers (positive, negative or null), whereas in the case of a "cos/sin" breakdown, the ranks are always natural integers (positive or null). If the signal $x(t)$ is real (i.e. a function of $\mathbb{R}$ in $\mathbb{R}$)[2], the Fourier decomposition presents a property known as Hermitian symmetry, which consists of noting that:

$$\forall k \in \mathbb{Z},\ c_{-k} = c_k^* \qquad \text{[A2.10]}$$

2 This is the most common situation encountered in power electronics.

A2.1.2.2. *Integration / derivation*

The integration $\int x(t).dt$ and the derivation $\dot{x}(t)$ of a signal $x(t)$ with a known Fourier series decomposition allow immediate calculation of the Fourier series of $\int x(t).dt$ and $\dot{x}(t)$. If we consider the following complex decomposition of $x(t)$:

$$x(t) = \sum_{k=-\infty}^{+\infty} c_k.\mathrm{e}^{j2k\pi Ft} \qquad [A2.11]$$

we deduce:

$$\int x(t).dt = \sum_{k=-\infty}^{+\infty} \underbrace{\frac{c_k}{j2k\pi F}}_{\gamma_k} \cdot \mathrm{e}^{j2k\pi Ft} \qquad [A2.12]$$

and:

$$\dot{x}(t) = \sum_{k=-\infty}^{+\infty} \underbrace{j2k\pi F.c_k}_{\delta_k} \cdot \mathrm{e}^{j2k\pi Ft} \qquad [A2.13]$$

A2.1.2.3. *Temporal dilation / contraction*

Temporal dilation or contraction consists of transforming an initial signal $x(t)$ into a signal $x(a.t)$ where $a \in \mathbb{R}^{+*}$. This type of transformation has no effect on the Fourier series decomposition or the way in which it is calculated. We must simply remember that the fundamental frequency $F = 1/T$ has been modified (along with the harmonics $k.F$), becoming $a.F$ (respectively, $a.k.F$).

### A2.1.3. *Parseval's theorem*

The RMS value $X_{\mathrm{rms}}$ of a signal $x(t)$ may be obtained by direct integration, applying the following definition:

$$X_{\mathrm{rms}}^2 = \frac{1}{T}\int_0^T x^2(t).dt \qquad [A2.14]$$

This is also possible using a link to coefficients $a_0$ and $c_k$ of the Fourier series:

$$X_{\text{rms}}^2 = a_0^2 + \sum_{k=1}^{+\infty} \left(a_k^2 + b_k^2\right) = \sum_{k=-\infty}^{+\infty} c_k . c_k^* \qquad \text{[A2.15]}$$

This formula is known as Parseval's theorem.

### A2.1.4. *Total harmonic distortion*

Parseval's theorem is extremely useful for calculating the total harmonic distortion (THD) of a non-sinusoidal value which we wish to compare to a pure sinusoid.

Two definitions of THD are used in two different standards:

– THD – F (IEEE or DIN standards) related to the fundamental of the value (which may be greater than 1):

$$\text{THD} - \text{F} = \frac{\sqrt{\sum_{k=2}^{+\infty} \left(a_k^2 + b_k^2\right)}}{\sqrt{a_1^2 + b_1^2}}; \qquad \text{[A2.16]}$$

– the TDH – F (IEC standard) related to the overall RMS value (always less than or equal to 1):

$$\text{THD} - \text{F} = \frac{\sqrt{\sum_{k=2}^{+\infty} \left(a_k^2 + b_k^2\right)}}{\sqrt{\sum_{k=1}^{+\infty} \left(a_k^2 + b_k^2\right)}}. \qquad \text{[A2.17]}$$

REMARK A2.1.– Note that, as a general rule, TDH – F (X = F or G) is calculated using quantities with a null continuous component (i.e. for $a_0 = 0$) or at least that the continuous component is not taken into account in calculating the THD.

## A2.2. Double Fourier series and PWM

### A2.2.1. *Context of study*

Before considering the spectral analysis of non-periodic signals, which will be covered in section A2.3, we will focus on one particularly important class of periodic signals encountered in power electronics: MLI signals, obtained by the modulation of a triangular (or sawtooth) carrier of frequency $F_d$ using a periodic modulation sequence (not necessarily sinusoidal) of frequency $F_m$. The effective determination of a Fourier series in this case is subject to significant calculation problems. The desired result may be obtained using a method based on a double Fourier series, proposed in [BEN 33] in 1933 and in [BLA 53] in 1953.

### A2.2.2. *Double Fourier series*

The double Fourier series is a generalization of the Fourier series to periodic functions of two variables of the type $f(x, y)$, with a period of $2\pi$ along the two axes[3]. As in the case of [A2.1], it is possible to write:

$$\begin{aligned} f(x,y) = A_{00} &+ \textstyle\sum_{n=1}^{\infty} \left(A_{0n}.\cos(ny) + B_{0n}.\sin(ny)\right) \\ &+ \textstyle\sum_{m=1}^{\infty} \left(A_{m0}.\cos(mx) + B_{m0}.\sin(mx)\right) \\ &+ \textstyle\sum_{n=1}^{\infty} \sum_{m=\pm 1}^{\pm\infty} \left(A_{mn}.\cos(mx+ny)\right. \\ &\left.+ B_{mn}.\sin(mx+ny)\right) \end{aligned} \qquad [A2.18]$$

where $\forall m \in \mathbb{N}, n \in \mathbb{Z}$:

$$A_{mn} = \frac{1}{2\pi^2} \int_0^{2\pi} \int_0^{2\pi} f(x,y).\cos(mx+ny).dx.dy \qquad [A2.19]$$

3 This does not limit the generality of the method, as this specific case can always be attained by changing a variable.

and:

$$B_{mn} = \frac{1}{2\pi^2} \int_0^{2\pi} \int_0^{2\pi} f(x, y) \,.\, \sin(mx + ny) \,.dx.dy \quad [A2.20]$$

### A2.2.3. *PWM and the "wall model"*

The "wall model" is based on the duplication of a time motif in the modulator across a series of vertical bands, with a width equal to the amplitude of the carrier, as shown in Figure A2.1. In this case, we will consider the simplest possibility using a sawtooth carrier: this type of carrier corresponds to an oblique line (AB) cutting across the copies, with regular modulator steps, for each switching instant of the PWM signal $c_{pwm}(t)$ for which we wish to calculate the spectrum. We must simply note that, when this line "travels through" the hatched zones, $c_{pwm} = 1$; in the white zones, $c_{pwm} = 0$.

Based on this representation, it is evidently possible to define a function $f(x, y)$ such that:

$$f(x, y) = \begin{cases} 1 \text{ in the hatched zone} \\ 0 \text{ otherwise} \end{cases} \quad [A2.21]$$

This function is periodic along both axes by construction: it can, therefore, be decomposed to produce a double Fourier series, as described in section A2.2.2. Finally, we must simply analyze the spectrum obtained along the line (AB) to obtain the RMS spectrum of the PWM signal. To do this, we note the relationship between variable $x = \omega_m t$ and $y = \omega_d t$[4] to obtain the desired result.

4 With these two variables $x$ and $y$, function $f(x, y)$ is $2\pi$-periodic along both axes.

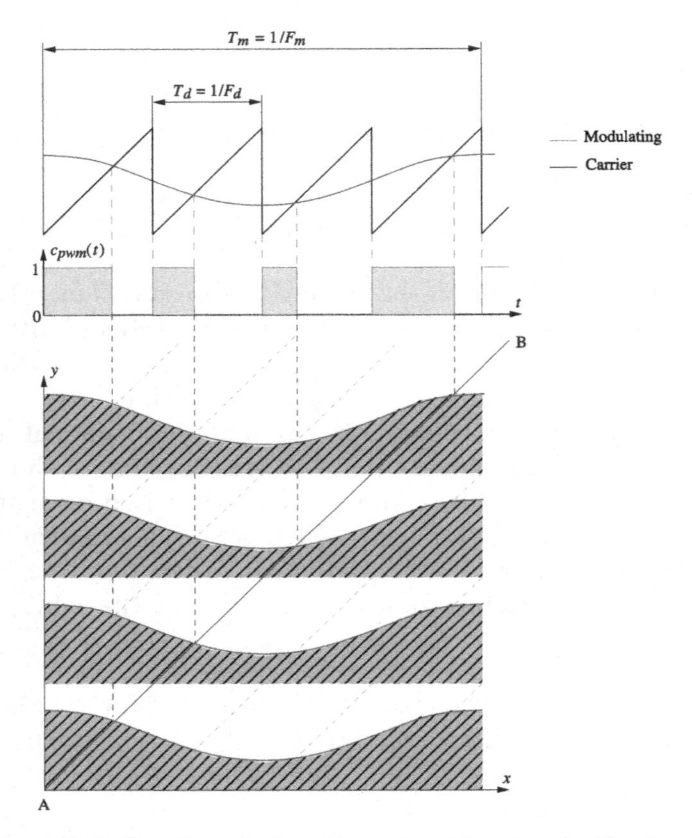


**Figure A2.1.** *Creation of a two-dimensional function for Fourier series decomposition. For a color version of the figure, see www.iste.co.uk/patin/power2.zip*

REMARK A2.2.– This particularly elegant method avoids the (major) difficulty of direct calculation of a PWM spectrum. However, the calculation is still relatively cumbersome, and involves Bessel functions of the first time, which can only be calculated approximately. We then simply read the curves produced by this family of functions to obtain an exact spectral representation of a given PWM signal.

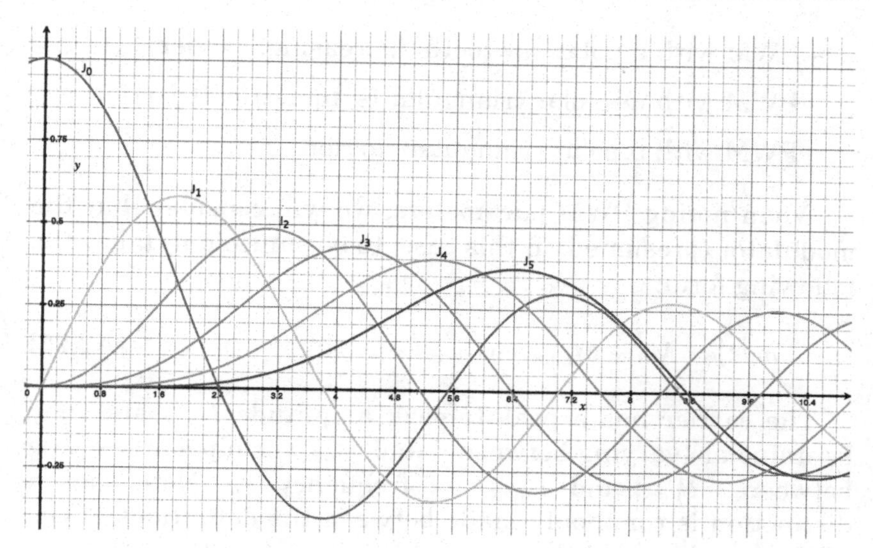


**Figure A2.2.** *Bessel functions of the first kind of order 0–5. For a color version of the figure, see www.iste.co.uk / patin / power2.zip*

## A2.2.4. *Bessel functions*

The Bessel function of the first kind of order $n$ is defined by the general formula:

$$J_n(x) = \frac{j^{-n}}{2\pi} \int_0^{2\pi} e^{jx\cos(\alpha)} e^{jn\cdot\alpha} d\alpha \qquad \text{[A2.22]}$$

The effective determination of the value of a Bessel function for any given argument is generally based on numerical calculations, or by reading the curves presented in Figure A2.2.

## A2.2.5. *Analytical spectra for different PWMs*

In this section, we will consider the Fourier series decompositions of a number of widespread PWM signal types:

– PWM with unipolar sawtooth (increasing) carrier;

– PWM with bipolar sawtooth (increasing) carrier;

– PWM with unipolar triangular (symmetrical) carrier;

– PWM with bipolar triangular carrier.

When noting PWM parameters, we will consider that the modulator (presumed, in this case, to be sinusoidal) takes the following form:

$$m(t) = M_0 + M_{\max}.\cos(\omega_m t + \phi_{m0}) \qquad \text{[A2.23]}$$

The carrier has a peak-to-peak amplitude $P_{\max}$ and may be unipolar (varying between 0 and $P_{\max}$) or bipolar (varying between $-P_{\max}$ and $P_{\max}$), of frequency $F_d$. Note that a distinction is commonly made between the definitions of the PWM signal $c_{pwm}$ in these two cases:

– for unipolar PWM, $c_{pwm} \in \{0, 1\}$;

– for bipolar PWM, $c_{pwm} \in \{-1/2, 1/2\}$ (but this may also be $\{-1, 1\}$).

Using these bases, we can define an average PWM signal $c_0 = \frac{M_0}{P_{\max}}$, whatever the strategy (unipolar or bipolar) and the depth of modulation $m = \frac{2M_{\max}}{P_{\max}}$ (once again, this is independent of the modulation type).

REMARK A2.3.– Using PWM signals with a unitary peak-to-peak amplitude for unipolar and bipolar PWM strategies makes it easier to carry out comparisons.

Furthermore, the instantaneous phase of the carrier is denoted as $\phi_p(t) = \omega_p t + \phi_{p0}$ where $\omega_p = 2\pi F_p$.

More information on PWM spectra may be found in [BLA 53], which also covers the case of decreasing sawtooth carriers.

A2.2.5.1. *PWM with a unipolar sawtooth carrier*

In this case, the signal $c_{pwm}$ is decomposed as follows:

$$c_{pwm}(t) = c_0 + \frac{m}{2}\cos(\omega_m t + \phi_{m0})$$
$$+\sum_{k=1}^{\infty}\frac{1}{k\pi}\left[\sin(k(\omega_p t + \phi_{p0}))\right.$$
$$\left.-J_0(km\pi).\sin(k(\omega_p t + \phi_{p0}) - 2kc_0\pi)\right]$$
$$+\sum_{k=1}^{\infty}\sum_{l=\pm1}^{\pm\infty}\frac{J_l(km\pi)}{k\pi}\sin\left(\frac{l\pi}{2} - k(\omega_p t + \phi_{p0})\right.$$
$$\left.-l(\omega_m t + \phi_{m0}) + 2kc_0\pi\right) \qquad [A2.24]$$

where we usually have $c_0 = 1/2$.

A2.2.5.2. *PWM with a bipolar sawtooth carrier*

In this case, we have (for $c_0 = 0$):

$$c_{pwm}(t) = \frac{m}{2}\cos(\omega_m t + \phi_{m0})$$
$$+\sum_{k=1}^{\infty}\frac{1}{k\pi}\left[\cos(k\pi) - J_0(km\pi).\sin(k(\omega_p t + \phi_{p0}))\right]$$
$$+\sum_{k=1}^{\infty}\sum_{l=\pm1}^{\pm\infty}\frac{J_l(km\pi)}{k\pi}\sin\left(\frac{l\pi}{2} - k(\omega_p t + \phi_{p0})\right.$$
$$\left.-l(\omega_m t + \phi_{m0})\right) \qquad [A2.25]$$

A2.2.5.3. *PWM with a unipolar triangular carrier*

Using this new carrier, we obtain:

$$c_{pwm}(t) = c_0 + \frac{m}{2}\cos(\omega_m t + \phi_{m0})$$
$$+\sum_{k=1}^{\infty}\frac{2}{k\pi}\cdot J_0\left(\frac{km\pi}{2}\right)\cdot\sin(k\pi c_0)\cos(k(\omega_p t + \phi_{p0}))$$

$$+\sum_{k=1}^{\infty}\sum_{l=\pm 1}^{\pm\infty}\frac{2}{k\pi}\cdot J_l\left(\frac{km\pi}{2}\right)$$

$$\cdot\sin\left(\frac{(2kc_0+l)\,\pi}{2}\right)\cos\left(k\left(\omega_p t+\phi_{p0}\right)+l\left(\omega_m t+\phi_{m0}\right)\right)\quad\text{[A2.26]}$$

As in the case of PWM with a unipolar sawtooth carrier, we generally take $c_0 = 1/2$.

#### A2.2.5.4. *PWM with a bipolar triangular carrier*

In this final case, we obtain (for $c_0 = 0$):

$$c_{pwm}(t)=\frac{m}{2}\cos\left(\omega_m t+\phi_{m0}\right)$$

$$+\sum_{k=1}^{\infty}\frac{2}{k\pi}\cdot J_0\left(\frac{km\pi}{2}\right)\cdot\sin\left(\frac{k\pi}{2}\right)\cos\left(k\left(\omega_p t+\phi_{p0}\right)\right)$$

$$+\sum_{k=1}^{\infty}\sum_{l=\pm 1}^{\pm\infty}\frac{2}{k\pi}\cdot J_l\left(\frac{km\pi}{2}\right)$$

$$\cdot\sin\left(\frac{(k+l)\,\pi}{2}\right)\cos\left(k\left(\omega_p t+\phi_{p0}\right)+l\left(\omega_m t+\phi_{m0}\right)\right)$$

[A2.27]

#### A2.2.5.5. *Qualitative summary*

In practice, triangular carrier PWMs are less rich in harmonic components (at least around frequency $F_d$) than those using a sawtooth carrier. Furthermore, unipolar PWM, for which we use voltage levels of 0 and $U_0$ during the positive alternations of the modulator and 0 and $-U_0$ for negative alternations, produces spectral content which is less rich than that produced by bipolar PWM (with the use of $\pm U_0$ over a switching period): this point is clearly illustrated in Chapter 2 of Volume 2 [PAT 15b].

## A2.3. Non-periodic signals

### A2.3.1. *Fourier transformation*

Fourier series can be extended to non-periodic signals using the notion of Fourier transformation. Any given signal $x(t)$ is associated with a Fourier transform $\mathcal{X}(f)$ defined as follows:

$$\mathcal{X}(f) = \mathcal{F}[x(t)] = \int_{\mathbb{R}} x(t).\mathrm{e}^{-j2\pi ft}.dt \qquad \text{[A2.28]}$$

### A2.3.2. *The Dirac impulse*

The unitary element of the Fourier transformation is a Dirac impulse $\delta(t)$. This is a distribution (generalization of mathematical functions) which can be assimilated, from a physical perspective, to a passage at the limit of a "gateway" function $\pi_T(t)$, defined as follows:

$$\pi_T(t) = \begin{cases} \frac{1}{T} \text{ pour } |t| \leq T/2 \\ 0 \text{ for } |t| > T/2 \end{cases} \qquad \text{[A2.29]}$$

Note, based on this definition, that:

$$\forall T \in \mathbb{R}^{+*}, \int_{\mathbb{R}} \pi_T(t).dt = 1 \qquad \text{[A2.30]}$$

The passage at the limit value leading to the Dirac impulse is thus:

$$\delta(t) = \lim_{T \to 0} \pi_T(t) \qquad \text{[A2.31]}$$

One important property of the Dirac impulse in the case of a function f(t) which is continuous in 0 is that:

$$\int_{\mathbb{R}} \delta(t).f(t).dt = f(0) \qquad \text{[A2.32]}$$

This enables us to establish the Fourier transform of the Dirac impulse $\Delta(f) = \mathcal{F}[\delta(t)]$:

$$\Delta(f) = 1 \qquad \text{[A2.33]}$$

Thus, we see that the spectral content of this impulse is uniform, and the spectral range is infinite. This is simply a mathematical tool, which has no physical reality in absolute terms: a Dirac impulse is impossible to obtain in practice, but remains useful for simplified modeling of brief events (which may be considered to be instantaneous[5] for the purposes of initial analysis).

### A2.3.3. *Properties*

#### A2.3.3.1. *Linearity*

As the Fourier transformation is an integral, its linearity is easy to verify:

$$\forall(\lambda,\mu)\in\mathbb{R}^2, \mathcal{F}[\lambda.p(t)+\mu.q(t)] = \lambda.\mathcal{F}[p(t)]+\mu.\mathcal{F}[q(t)] \qquad \text{[A2.34]}$$

#### A2.3.3.2. *Integration / derivation*

Let us consider a signal $x(t)$, with a known Fourier transform denoted as $\mathcal{X}(f) = \mathcal{F}[x(t)]$. We will begin by establishing the expression of the Fourier transform of $\int_{-\infty}^{t} x(\tau).d\tau$:

$$\mathcal{F}\left[\int_{-\infty}^{t} x(\tau).d\tau\right] = \int_{\mathbb{R}}\int_{-\infty}^{t} x(\tau).d\tau.\mathrm{e}^{-j2\pi ft}.dt \qquad \text{[A2.35]}$$

REMARK A2.4.– The formula for integration by parts can be deduced from the product derivation formula:

$$(uv)' = u'v + uv' \qquad \text{[A2.36]}$$

5 Switching, for example, in the context of power electronics.

This gives us the following result:

$$\int uv' = [uv] - \int u'v \qquad \text{[A2.37]}$$

Using [A2.37], based on [A2.35], we obtain the following result:

$$\mathcal{F}\left[\int_{-\infty}^{t} x(\tau).d\tau\right] = \left[\int_{-\infty}^{t} x(\tau).d\tau \cdot \frac{e^{-j2\pi ft}}{-j2\pi f}\right]_{-\infty}^{+\infty} + \frac{1}{j2\pi f} \cdot \int_{\mathbb{R}} x(t).e^{-j2\pi ft}.dt \qquad \text{[A2.38]}$$

Supposing that the integrated function tends toward zero toward infinity (i.e. in $\pm\infty$), the first term disappears. Hence:

$$\mathcal{F}\left[\int_{-\infty}^{t} x(\tau).d\tau\right] = \frac{1}{j2\pi f} \cdot \int_{\mathbb{R}} x(t).e^{-j2\pi ft}.dt = \frac{\mathcal{X}(f)}{j2\pi f} \qquad \text{[A2.39]}$$

For derivation, we wish to calculate the Fourier transform of $\dot{x}(t) = \frac{dx}{dt}$:

$$\mathcal{F}[\dot{x}(t)] = \int_{\mathbb{R}} \dot{x}(t).e^{-j2\pi ft}.dt \qquad \text{[A2.40]}$$

Similarly to the case of integration, we can establish the following relationship (using integration by parts):

$$\mathcal{F}[\dot{x}(t)] = j2\pi f.\mathcal{X}(f) \qquad \text{[A2.41]}$$

A2.3.3.3. *Temporal dilatation / contraction*

The problem of temporal dilation and contraction for the Fourier transform is different from that encountered using Fourier series for periodic signals. However, the starting point for study still consists of replacing a signal $x(t)$ with a known Fourier transform $\mathcal{X}(f)$ by a signal $x(a.t)$ with a strictly positive real coefficient $a$ (i.e. $a \in \mathbb{R}^{+*}$). We may begin by defining the Fourier transform of the new signal:

$$\mathcal{F}\left[x\left(a.t\right)\right] = \int_{\mathbb{R}} x\left(a.t\right).\mathrm{e}^{-j2\pi ft}.dt \qquad \text{[A2.42]}$$

We then simply change a variable ($\tau = a.t$ and thus $t = \frac{\tau}{a}$) to obtain a result. First, note that $dt = \frac{d\tau}{a}$; as $a > 0$, integration is always carried out from $-\infty$ to $+\infty$ (and not in the opposite direction). Hence:

$$\mathcal{F}\left[x\left(a.t\right)\right] = \frac{1}{a}\mathcal{X}\left(\frac{f}{a}\right) \qquad \text{[A2.43]}$$

A2.3.3.4. *Hermitian symmetry*

Hermitian symmetry, as seen in the context of complex Fourier series, also occurs in the case of the Fourier transform. When the signal $x(t)$ under study is real, symmetry will be present between the value of the Fourier transform $\mathcal{X}(f)$ in $f$ and in $-f$. This is why the representation of a signal spectrum is generally limited to a unilateral representation for $f \geq 0$, and not to the bilateral form, which provides no additional information. To demonstrate this symmetry, note the expression of the Fourier transform of $x(t)$:

$$\mathcal{X}(f) = \int_{\mathbb{R}} x(t).\mathrm{e}^{-j2\pi ft}.dt \qquad \text{[A2.44]}$$

The conjugation operation (denoted as$\mathrm{conj}(z) = z^*$) is linear and can consequently be placed inside or outside of the $\int$ sign. Thus:

$$\mathcal{X}(f)^* = \left(\int_{\mathbb{R}} x(t).\mathrm{e}^{-j2\pi ft}.dt\right)^* = \int_{\mathbb{R}} \left(x(t).\mathrm{e}^{-j2\pi ft}\right)^*.dt \qquad \text{[A2.45]}$$

The conjugation of a product is equal to the product of the conjugations:

$$\mathcal{X}(f)^* = \int_{\mathbb{R}} x(t)^*.e^{j2\pi ft}.dt \qquad \text{[A2.46]}$$

If $x(t)$ is real, we have $x(t) = x(t)^*$ and thus:

$$\mathcal{X}(f)^* = \int_{\mathbb{R}} x(t).e^{j2\pi ft}.dt = \mathcal{X}(-f) \qquad \text{[A2.47]}$$

This result is known as the Hermitian symmetry of a Fourier transform (of a real signal), and corresponds to the continuous form of result [A2.10], obtained for complex Fourier series.

A2.3.3.5. *Time reversal*

We have already considered the impact of temporal dilation/contraction on the Fourier transform of a signal, using a strictly positive temporal modification coefficient ($a > 0$). We may also wish to consider the case where $a$ is negative (non-null), or the specific case where signal $x(t)$ is replaced by an opposite signal in relation to the time axis $x(-t)$. Note that the composition of the two effects produces a general case, corresponding to $a \in \mathbb{R}^*$:

$$\mathcal{F}[x(-t)] = \int_{\mathbb{R}} x(-t).e^{-j2\pi ft}.dt \qquad \text{[A2.48]}$$

Once again, we must change a variable ($\tau = -t$, and thus $dt = -d\tau$). Note that, in this case, the direction of integration is also reversed:

$$\mathcal{F}[x(-t)] = -\int_{+\infty}^{-\infty} x(\tau).e^{j2\pi f\tau}.d\tau = \int_{-\infty}^{+\infty} x(\tau).e^{j2\pi f\tau}.d\tau$$
$$= \mathcal{X}(-f) \qquad \text{[A2.49]}$$

Based on the Hermitian symmetry result established in the previous section, we obtain:

$$\mathcal{F}\left[x\left(-t\right)\right] = \mathcal{X}\left(f\right)^{*} \qquad \text{[A2.50]}$$

A2.3.3.6. *Lag*

When a signal $x(t)$ with a known Fourier transform $\mathcal{X}(f)$ is delayed for a duration $t_0$, we can easily verify, by changing a variable, that:

$$\mathcal{F}\left[x\left(t - t_0\right)\right] = \mathcal{X}\left(f\right).\mathrm{e}^{-j2\pi f t_0} \qquad \text{[A2.51]}$$

A2.3.3.7. *Frequency translation*

When we multiply a signal $x(t)$ with a known Fourier transform $\mathcal{X}(f)$ by a complex exponential $\mathrm{e}^{j2\pi f_0 t}$, it is possible to show that the convolution in terms of frequency leads to a frequency translation:

$$\mathcal{F}\left[x\left(t\right).\mathrm{e}^{j2\pi f_0 t}\right] = \mathcal{X}\left(f - f_0\right) \qquad \text{[A2.52]}$$

A2.3.3.8. *Convolution*

The convolution product $\star$ is a mathematical operation which is widely used in physics in relation to the solution of ordinary differential equations (the tool may also be generalized for the solution of partially derived equations, with the addition of a Green node). Unfortunately, this operation is hard to process directly in practice, as it concerns the "sliding" integral of the product of two functions between $-\infty$ and $+\infty$, as demonstrated by the following definition:

$$r(t) = \left(p \star q\right)(t) = \int_{\mathbb{R}} p\left(\tau\right).q\left(t - \tau\right).d\tau \qquad \text{[A2.53]}$$

However, this operation may be carried out in a simplified manner in an image domain: the Fourier (frequency) domain, and more generally the Laplace domain (used in automatics, and discussed in Volume 3 [PAT 15c], Chapter 4, in the context of switch-mode power supply transfer), may be treated by introducing the Laplace variable $p$. The simplification operation consists of noting that the convolution product becomes a simple product (i.e. an arithmetic multiplication) in this image domain.

Let us consider two signals $p(t)$ and $q(t)$, with Fourier transforms denoted as $\mathcal{P}(f)$ and $\mathcal{Q}(f)$, respectively. The product of convolution $r(t)$ between $p(t)$ and $q(t)$, defined in accordance with equation [A2.53], presents a Fourier transform $\mathcal{R}(f)$ which may be expressed using the following formula:

$$\mathcal{R}(f) = \mathcal{P}(f) . \mathcal{Q}(f) \tag{A2.54}$$

Note that, while this result appears to be simple, a (potentially considerable) difficulty remains concerning the return to the temporal domain. To do this, we need to use an inverse Fourier transformation formula, and we must be able to apply this formula to the result obtained in [A2.54].

In the case where one of the two functions is replaced by the Dirac impulse, which is the neutral element of the convolution product (here, using any given function $f(t)$), calculation is simple:

$$(f \star \delta)(t) = (\delta \star f)(t) = f(t) \tag{A2.55}$$

This is also valid in the frequency domain, as the Fourier transform $\Delta(f)$ of the Dirac impulse has a value of 1, as demonstrated in [A2.33].

REMARK A2.5.– The Fourier transform allows us to replace the convolution product by a simple product, as shown above, but the reverse is also true. A simple product may be replaced by a convolution product using a Fourier transformation.

A2.3.3.9. *Inverse transformation*

The inverse Fourier transformation will be defined below. In this case, consider a Fourier transform $\mathcal{X}(f)$ from which we wish to obtain the temporal original $x(t)$:

$$x(t) = \int_{\mathbb{R}} \mathcal{X}(f) . e^{j2\pi ft} . df \qquad \text{[A2.56]}$$

REMARK A2.6.– The inverse Fourier transformation formula is very similar to the direct transformation formula, and its properties are similar, notably in relation to the convolution product.

A2.3.3.10. *Sinusoids*

The Fourier transform of $\cos(2\pi f_0 t)$ can be obtained using Euler's formula:

$$\cos(2\pi f_0 t) = \frac{e^{j2\pi f_0 t} + e^{-j2\pi f_0 t}}{2} \qquad \text{[A2.57]}$$

Consequently, the Fourier transformation gives us the following result:

$$\mathcal{F}[\cos(2\pi f_0 t)] = \frac{1}{2} \int_{\mathbb{R}} \left( e^{-j2\pi(f-f_0)t} + e^{-j2\pi(f+f_0)t} \right) . dt \qquad \text{[A2.58]}$$

A useful result consists of noting, based on [A2.33] and [A2.56], that:

$$\int_{\mathbb{R}} e^{j2\pi ft} . df = \delta(t) \qquad \text{[A2.59]}$$

In the same way, as $\delta(t)$ is even, we may also write:

$$\int_{\mathbb{R}} e^{-j2\pi ft} . df = \delta(t) \qquad \text{[A2.60]}$$

Note also that the roles of $t$ and $f$ are completely interchangeable in these results.

Consequently, after changing the variable, we obtain:

$$\mathcal{F}\left[\cos\left(2\pi f_0 t\right)\right] = \frac{1}{2}\left(\delta\left(f - f_0\right) + \delta\left(f + f_0\right)\right) \qquad [A2.61]$$

In the case of the signal $\sin\left(2\pi f_0 t\right)$, we begin by noting:

$$\sin\left(2\pi f_0 t\right) = \frac{\mathrm{e}^{j2\pi f_0 t} - \mathrm{e}^{-j2\pi f_0 t}}{2j} \qquad [A2.62]$$

We then deduce the spectrum, as in the case of [A2.61]:

$$\mathcal{F}\left[\cos\left(2\pi f_0 t\right)\right] = \frac{1}{2j}\left(\delta\left(f - f_0\right) - \delta\left(f + f_0\right)\right) \qquad [A2.63]$$

REMARK A2.7.– We see that Hermitian symmetry, as described in [A2.47], is respected for the two results [A2.61] and [A2.63].

### A2.3.4. *Fourier transform of periodic signals*

Using the Fourier transform of any periodic signal, we may expect to obtain a discrete spectrum (non-null only at multiples of the fundamental frequency), corresponding to the complex Fourier series decomposition.

Let us consider a signal $m_T(t)$ with a finite temporal support $T$ (i.e. non-null for an interval of width $T$ alone). It is interesting to note that the convolution of this signal by a delayed Dirac impulse $\delta_{t_0}(t) = \delta\left(t - t_0\right)$ gives us a delayed version of the signal:

$$\left(m_T \star \delta_{t_0}\right)(t) = m_T\left(t - t_0\right) \qquad [A2.64]$$

Based on this result, a $T$-periodic signal $m(t)$ may be formed using the motif $m_T(t)$ using convolution between this initial signal and a Dirac comb (sampling function) of period $T$:

$$\perp_T(t) = \sum_{k \in \mathbb{Z}} \delta\left(t - kT\right) \qquad [A2.65]$$

Hence:

$$m(t) = (m_T \star \perp_T(t)) \qquad [A2.66]$$

Next, if we wish to calculate the Fourier transform $\mathcal{M}(f)$ of the obtained signal, we have:

$$\mathcal{M}(f) = \mathcal{M}_T(f) . \mathcal{F}\left[\sum_{k\in\mathbb{Z}} \delta(t - kT)\right] \qquad [A2.67]$$

As the Fourier transformation is a linear operation, it may be applied to each element under the $\sum$ sign separately:

$$\mathcal{M}(f) = \mathcal{M}_T(f) . \sum_{k\in\mathbb{Z}} \mathcal{F}[\delta(t - kT)] \qquad [A2.68]$$

Using [A2.33] and [A2.51], we have:

$$\mathcal{F}[\delta(t - kT)] = \mathrm{e}^{-j2k\pi fT} \qquad [A2.69]$$

Hence:

$$\mathcal{M}(f) = \mathcal{M}_T(f) . \sum_{k\in\mathbb{Z}} \mathrm{e}^{-j2k\pi fT} \qquad [A2.70]$$

The (frequency) periodicity $1/T$ of $X(f) = \sum_{k\in\mathbb{Z}} \mathrm{e}^{-j2k\pi fT}$ is easy to demonstrate. This spectrum may then be decomposed to produce a Fourier series. Moreover, this expression corresponds precisely to a complex decomposition in which all of the coefficients $c_k$ (for any relative integer $k$) have a value of 1. For a signal $x(t)$, coefficient $c_k$ is expressed as:

$$c_k = \frac{1}{T} \int_0^T x(t) . \mathrm{e}^{j2k\pi Ft} . dt \qquad [A2.71]$$

In the case of our frequency (with period $1/T$), this corresponds to:

$$c_k = T.\int_0^{1/T} X(f).e^{j2k\pi Tf}.df \qquad \text{[A2.72]}$$

The integration interval should be of width $1/T$, but the boundaries may be modified: for example, we may choose an interval centered on $f = 0$ (between $-\frac{1}{2T}$ and $\frac{1}{2T}$). In this case, we wish to find a function $X(f)$ such that:

$$\forall k \in \mathbb{Z},\ c_k = 1 \qquad \text{[A2.73]}$$

It is easy to verify that:

$$X(f) = \frac{1}{T} \cdot \delta(f) \qquad \text{[A2.74]}$$

is a solution in the interval $\left[-\frac{1}{2T}; \frac{1}{2T}\right]$. Its existence is, therefore, proved, and in this case, unique. Consequently, the global expression of $X(f)$ (i.e. $\forall f \in \mathbb{R}$) is:

$$X(f) = \frac{1}{T} \cdot \sum_{k\in\mathbb{Z}} \delta\left(f - \frac{k}{T}\right) \qquad \text{[A2.75]}$$

We see that the spectrum of a temporal Dirac comb is a Dirac frequency comb. Expression [A2.70] of the spectrum of a periodic signal becomes:

$$\mathcal{M}(f) = \frac{1}{T} \cdot \mathcal{M}_T(f) . \sum_{k\in\mathbb{Z}} \delta\left(f - \frac{k}{T}\right) \qquad \text{[A2.76]}$$

We see that the continuous spectrum of the elementary motif $m_T(t)$ (defined over a single period $T$) is sampled at all multiples of the fundamental frequency $1/T$. This result is perfectly coherent with the expected discrete spectrum, and conforms to the complex Fourier series decomposition defined in [A2.6]–[A2.7].

### A2.3.5. *Fourier transform of sampled signals*

The result presented in the previous section has a counterpart associated with the Fourier transform of a sampled signal. It is possible to demonstrate that a sampled signal (with discrete temporal components) is associated with a periodic frequency spectrum.

To do this, we associate a signal $x(t)$ with its sampled version $x^*(t)$, obtained by multiplying $x(t)$ by a temporal Dirac comb of period $T$:

$$x^*(t) = x(t). \sum_{k\in\mathbb{Z}} \delta\left(t - kT\right) \qquad \text{[A2.77]}$$

The Fourier transform $X^*(f)$ of this signal is obtained by convolution of spectrum $X(f)$ of $x(t)$ by the spectrum of the temporal Dirac comb:

$$X^*\left(f\right) = X\left(f\right) \star \mathcal{F}\left[\sum_{k\in\mathbb{Z}} \delta\left(t - kT\right)\right] \qquad \text{[A2.78]}$$

As the Fourier transform is a linear operation, we have:

$$\mathcal{F}\left[\sum_{k\in\mathbb{Z}} \delta\left(t - kT\right)\right] = \sum_{k\in\mathbb{Z}} \mathcal{F}\left[\delta\left(t - kT\right)\right] = \sum_{k\in\mathbb{Z}} \mathrm{e}^{-j2k\pi fT} \qquad \text{[A2.79]}$$

Therefore, we can write:

$$X^*\left(f\right) = X\left(f\right) \star \sum_{k\in\mathbb{Z}} \mathrm{e}^{-j2k\pi fT} \qquad \text{[A2.80]}$$

Note that $\sum_{k\in\mathbb{Z}} \mathrm{e}^{-j2k\pi fT}$ also appeared in equation [A2.70], where it was identified as a Dirac frequency comb. In this case, the spectrum is convolved with $X(f)$, giving the following result:

$$X^*(f) = \sum_{k \in \mathbb{Z}} X\left(f - \frac{k}{T}\right) \qquad \text{[A2.81]}$$

The spectrum of the sampled signal is thus a duplication of the spectrum of the initial signal $x(t)$ around each multiple of the sampling frequency $1/T$. Note that if the spectrum $X(f)$ is bounded by a maximum frequency denoted as $f_{\max}$, the inequality $2f_{\max} \leq 1/T$ must be respected to avoid overlap in the duplicated motifs of spectrum $X(f)$. This inequality is known as Shannon's theorem and the overlap phenomenon is known as aliasing. This result is important not only in digital signal processing but also in explaining certain phenomena encountered in PWM, particularly the appearance of subharmonics when the switching frequency is too low in relation to the modulation frequency.

### A2.3.6. *Parseval's theorem*

As for Fourier series, Parseval's theorem is applicable to Fourier transformations. However, there is one important nuance in this case: in the case of periodic signals, the integration interval for the square of the signal is limited to the period, whereas in the context of the Fourier transformation, the integration interval covers the whole of the real axis. The treated function must, therefore, be of class $\mathcal{L}_2$ (i.e. a summable square function).

Let us take a signal of this type, $x(t)$ (presumed to be complex in this case to ensure generality). An identity exists between the temporal and frequency integrals:

$$\int_{\mathbb{R}} |x(t)|^2 .dt = \int_{\mathbb{R}} |\mathcal{X}(f)|^2 .df \qquad \text{[A2.82]}$$

### A2.3.7. *The Heisenberg–Gabor spectrum inequality*

The result presented here provides an important basis not only for signal theory, but also for quantum mechanics, where it is known as the Heisenberg uncertainty principle (established in 1927); Heisenberg was awarded the Nobel prize in physics in 1933 for creating this new area of research.

In qualitative terms, the result may be summarized as follows: a short signal (in temporal terms) occupies a broad range of frequencies. However, a signal which is highly localized in terms of frequency is longer in terms of time.

This result is clearly shown in two of the examples seen above:

– The Dirac impulse is the shortest possible signal, and, as we have seen, its spectrum is uniform up to $f \to \infty$.

– The sinusoid (with frequency $f_0$) is a signal with a spectrum (unilateral) limited to a single component at $f = f_0$. However, it occupies a time range from $-\infty$ to $+\infty$.

We will now consider the quantitative aspects of these statements. To do this, we need to introduce the notions of temporal and frequency dispersion.

#### A2.3.7.1. *Temporal dispersion*

The temporal dispersion $\sigma_t$ of a signal $\psi(t)$ is defined in a way similar to the standard deviation of a random signal:

$$\sigma_t = \left( \frac{\int (t - \bar{t})^2 . |\psi(t)|^2 .dt}{\int |\psi(t)|^2 .dt} \right)^{1/2} \quad \text{[A2.83]}$$

where $\bar{t}$ is the temporal barycenter of the signal:

$$\bar{t} = \int t.\psi(t) .dt \quad \text{[A2.84]}$$

A2.3.7.2. *Frequency dispersion*

A similar approach is used to calculate the frequency dispersion (or, more correctly, the dispersion of the angular frequency $\omega = 2\pi f$):

$$\sigma_\omega = \left( \frac{\int (\omega - \overline{\omega})^2 . |\Psi(\omega)|^2 .d\omega}{\int |\Psi(\omega)|^2 .d\omega} \right)^{1/2} \quad \text{[A2.85]}$$

where $\Psi(\omega = 2\pi.f) = \mathcal{F}[\psi(t)]$ with a frequency barycenter $\overline{\omega}$ defined as follows:

$$\overline{\omega} = \int \omega.\Psi(\omega).d\omega \quad \text{[A2.86]}$$

A2.3.7.3. *Heisenberg–Gabor inequality*

This inequality, applied to any given signal $\psi(t)$, may be summarized as:

$$\sigma_t.\sigma_\omega \geq \frac{1}{2} \quad \text{[A2.87]}$$

REMARK A2.8.– The demonstration of this inequality lies outside the scope of this book, but further details may be found in [DEG 01].

### A2.3.8. *"Time/frequency" optimal signal*

We may use inequality [A2.87] to consider the form of the signal $\psi(t)$ which allows us to reach a situation of equality, which may be considered to be optimal. It is possible to verify that this result is obtained for a Gaussian signal $g(t)$:

$$g(t) = \frac{1}{\sigma_t\sqrt{2\pi}}.e^{-\frac{t^2}{2\sigma_t^2}} \quad \text{[A2.88]}$$

Note that this signal is centered on instant $t = 0$, and the temporal dispersion $\sigma_t$ appears explicitly in the expression.

Moreover, the Fourier transform of a Gaussian signal is also Gaussian:

$$G(\omega) = e^{-\sigma_t^2 \omega^2} \quad \text{[A2.89]}$$

Using the analogy between the two expressions [A2.88] and [A2.89], the angular frequency distribution $\sigma_\omega$ may be obtained without calculation:

$$\sigma_\omega = \frac{1}{2\sigma_t} \quad \text{[A2.90]}$$

We, therefore, clearly see that the minimum boundary of the product $\sigma_t.\sigma_\omega$ is reached in this case.

## A2.4. PWM and distortion analysis

In this section, we will consider the quality of the power supply to a load in permanent sinusoidal load provided by an inverter (in both single- and three-phase contexts). In this case, we will presume that the inverter and power supply are ideal:

– a strictly constant voltage source entering the inverter;

– instantaneous switch commutation;

– no deadtime in switching in half-bridges;

– zero voltage drop-off at the switch terminals in ON state.

Evidently, the voltage $v(t)$ supplied to the load has a finite number of possible values, due to the switching function of the converter used in the power supply. This value is a piecewise constant (i.e. constant for given time intervals). However, using PWM, the sliding average of this voltage needs to follow a reference sinusoid with fixed amplitude and frequency values. The load, generally of the R, L, E type for an electrical machine, behaves as a low-pass filter which eliminates (or at least significantly limits) the high-frequency components of the current. This specific value is key in evaluating the quality of the power supply to a machine, as it is central to the torque generated by the machine in relation to the mechanical load.

Since a machine is a complex piece of equipment, the characterization of a PWM strategy in terms of distortion requires the use of mathematical tools, which must be sufficiently representative of the real load and sufficiently simple to enable effective study. Generally, an evaluation of the integral of the voltage wave is used to evaluate the distortion created by a command in comparison with both the ideal case and a number of other command techniques.

This methodology is applicable to both single- and three-phase inverters. However, in the case of a three-phase inverter, we use a vector-based approach to model the inverter, with an equivalent two-phase representation of the voltage (and current) output of the inverter. Despite this difference, we will systematically consider an inverter output voltage waveform $v(t)$, in comparison with the desired sinusoidal wave $v_{\mathrm{ref}}(t)$. The induced error, denoted as $\Delta(t) = v(t) - v_{\mathrm{ref}}(t)$, is then integrated to obtain a signal denoted as $\Sigma(t) = \int \Delta(t).dt$. We then evaluate the RMS value of this signal over a period $F_m$ of the reference wave $v_{\mathrm{ref}}(t)$.

Clearly, a certain number of additional parameters have an effect on the result:

– the direct current (DC) bus voltage $V_{\mathrm{dc}}$ powering the inverter;

– the amplitude of the reference voltage $V_{\mathrm{ref}}^{\max}$;

– the switching frequency $F_d$.

In practice, these parameters may be condensed to give two normalized parameters:

– the modulation depth $K_m = 2V_{\text{ref}}^{\max}/V_{\text{dc}}$;

– the frequency ration $K_f = F_d/F_m$.

### A2.4.1. *Single-phase inverters*

In a single-phase context, only two fixed-frequency command variations may be envisaged for an H-bridge, single-phase inverter. Two types of PWM may be used:

– bipolar PWM (with complementary control of the two half-bridges, giving a voltage of $v(t) \in \{-V_{\text{dc}}; +V_{\text{dc}}\}$);

– unipolar PWM (with one half-bridge in the ON state for each half-alternation of $v_{ref}(t)$, giving a voltage of $v(t) \in \{-V_{\text{dc}}; 0; +V_{\text{dc}}\}$).

In comparing the two strategies, we have chosen to use the THD of the integral of the error between $v(t)$ and $v_{ref}(t)$ ($v_{ref}$ is sinusoidal, with a period $T_m = 1/F_m$):

$$\text{THD}_{weighted} = \frac{\sqrt{\frac{1}{T_m}\int_0^{T_m}\left(\int_0^t v(\tau) - v_{ref}(\tau)d\tau\right)^2 .dt}}{V_{\text{ref}}^{\max}/\sqrt{2}} \quad \text{[A2.91]}$$

The logarithm of this distortion rate ($\log(\text{THD}_{pond})$) for both modulation types is presented in Figure A2.3.

We immediately see that the distortion resulting from unipolar PWM is significantly lower for a given pairing $(K_f, K_m)$ than the distortion involved in bipolar PWM. In qualitative terms, this result can be explained by the fact that the voltage peaks induced by unipolar PWM are half the size of those induced by bipolar PWM ($V_{\text{dc}}$ instead of $2V_{\text{dc}}$), as shown in Chapter 2 of Volume 2 [PAT 15b].

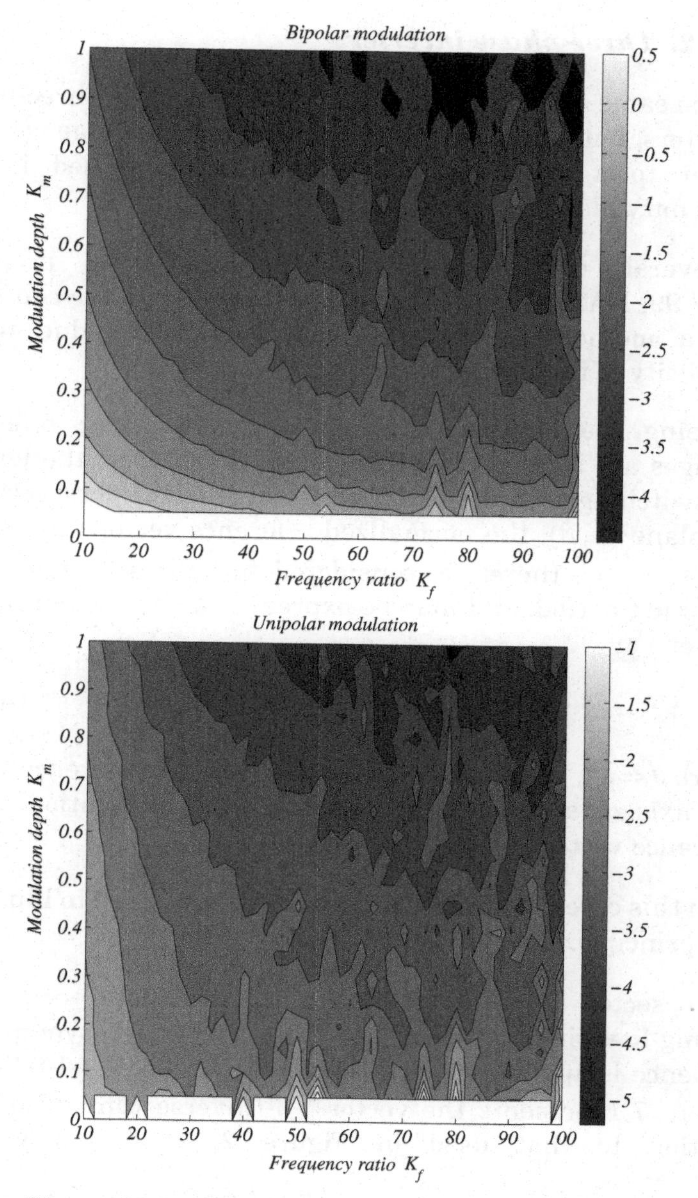


**Figure A2.3.** *Comparison of distortions resulting from bipolar and unipolar PWM*

### A2.4.2. *Three-phase inverters*

The same analytical approach may be used for three-phase inverters, but it no longer applies to a scalar voltage $v(t)$, but rather to a two-phase voltage vector, expressed in the stationary plane $(\alpha, \beta)$[6].

Several calculation methods have been proposed [HAV 99, NAR 08, NAR 06, ZHA 10]; here, we have chosen to use a method developed by Hava [HAV 98] due to its simplicity of implementation.

Using the hypothesis $F_d \gg F_m$, the three reference voltages are always considered to be constant at the level of the switching period $T_d = 1/F_d$. From a vector perspective in the plane $(\alpha, \beta)$, the normalized reference vector (in relation to $\frac{V_{dc}}{2}$) $\overrightarrow{V^*}$ is, therefore, considered to be constant for each switching period, and may be expressed using the equivalent phaser $\overline{V^*}$:

$$\overline{V^*} = m \times e^{j\theta} \qquad \text{[A2.92]}$$

where $\theta = \omega t$, which is the angle between the reference vector and axis $\alpha$, and $\omega$ is the angular speed of rotation of the reference vector.

In this case, the SVPWM strategy may be used to illustrate this principle.

In sector I of the hexagon in the plane $\alpha\beta$, for a raising-lowering-type carrier, the following symmetrical sequence is applied for each switching period: 7-2-1-0-0-1-2-7 (where 7 represents the vector $\overrightarrow{V_7}$) (inverse configuration in relation to that used in Figure 2.15 of this volume,

6 This approach can also be applied to quantities expressed in a rotating plane $(d, q)$: both formalisms are used in the literature on the subject, but note that the results obtained are strictly equivalent.

Chapter 2). When a normalized output vector of the inverter $\overrightarrow{V_i}$ is applied, an instantaneous error vector (or harmonic vector) is deduced using the following relationship:

$$\overrightarrow{\Delta_i} = \frac{v_{\mathrm{dc}}}{2} \times \underbrace{(\overrightarrow{V_i} - \overrightarrow{V^*})}_{\overrightarrow{\delta_i}} \qquad \text{[A2.93]}$$

where $\overrightarrow{\delta_i}$, dependent on $m$, $\theta$ and $\overrightarrow{V_i}$, is the normalized vector in relation to $V_{\mathrm{dc}}/2$ of vector $\overrightarrow{\Delta_i}$.

This voltage error $\overrightarrow{\Delta_i}$ is, evidently, measured in volts, and vector $\overrightarrow{\delta_i}$ has no unit.

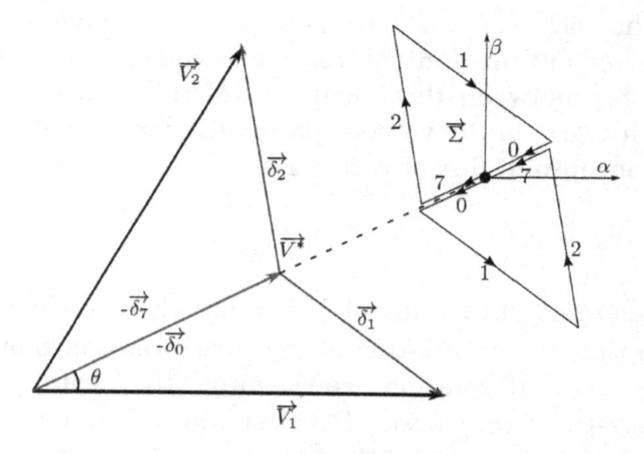


**Figure A2.4.** *Trajectory of the harmonic flux during a switching period for the SVPWM strategy*

Figure A2.4 shows the different error vectors corresponding to a reference vector situated in sector I of the hexagon. We see that the amplitude and phase of the error vectors are dependent on the amplitude (and thus the modulation index $m$) and the position of the reference vector (and thus on angle $\theta$).

The cumulated voltage error is defined by the following formula:

$$\overrightarrow{\Sigma} = \int \overrightarrow{\Delta_i} dt = \frac{v_{dc}}{2} \times \int \overrightarrow{\delta_i} dt \quad \text{[A2.94]}$$

This quantity $\overrightarrow{\Sigma}$ is the integral of a voltage, and is equivalent (following Faraday's law) to a magnetic flux. This flux is known as the conceptual harmonic flux.

Using a classic $R - L(-E)$ model of electrical machines (this time using three phases), the role of the inductive component is more important than that of the resistive component at switching period level. As the origin of the current harmonics is the error between the voltage applied at the inverter output and the reference voltage, the following relationship between the harmonics of the load currents $I_h$ and the integral of the voltage error vector reveals the nature of the "conceptual" flux of vector $\Sigma$:

$$\overrightarrow{\Sigma} = L \times \overrightarrow{I_h} \quad \text{[A2.95]}$$

Consequently, the study of $\overrightarrow{\Sigma}$ is equivalent to the study of $\overrightarrow{I_h}$. Note that the calculation of the conceptual harmonic flow requires no information concerning the load, and is characteristic of the chosen PWM strategy. The trajectory of $\overrightarrow{\Sigma}$ corresponding to the SVPWM strategy over a switching period is illustrated in Figure A2.4. Let us suppose that, at the start of the first switching period, $\overrightarrow{\Sigma}$ starts from 0: it returns to 0 in the middle and at the end of the switching period (this is repeated for all periods). The trajectory shown in Figure A2.4 corresponds to the 7-2-1-0-0-1-2-7 sequence, with equal application times for vectors $\overrightarrow{V_0}$ and $\overrightarrow{V_7}$. As intersective PWM strategies with raising lowering-type carriers only generate symmetrical switching sequences, it is sufficient to calculate $\overrightarrow{\Sigma}$ for half of the switching period, and the trajectory of $\overrightarrow{\Sigma}$ for the second half of the period is exactly symmetrical to that of the first half. Furthermore, each PWM strategy involves a different distribution of the application times of free wheel vectors; consequently, the trajectory of $\overrightarrow{\Sigma}$ for each PWM strategy is unique (we do not, therefore, obtain results by PWM family, as in the case of $\overline{I_{dc}}$).

We will now present the method used to calculate the RMS value of the modulus of $\overrightarrow{\Sigma}$ over a fundamental period; symmetry in the plane $\alpha\beta$ means that only a sector of $\frac{\pi}{3}$ (60°) needs to be analyzed.

First, variable changes may be used to express the harmonic flux over half a switching period as:

$$\overrightarrow{\Sigma} = \Sigma_0 \times \overrightarrow{\sigma} \qquad \text{[A2.96]}$$

where $\Sigma_0 = \frac{v_{dc}}{2}\frac{T_d}{2}$ is dependent on the DC bus voltage and the switching period, and $\overrightarrow{\sigma}$ is the normalized vector in relation to $\Sigma_0$ of vector $\overrightarrow{\Sigma}$.

For the SVPWM strategy, the analytical formulation of the trajectory of phaser $\overline{\sigma}$ (associated with vector $\sigma$) is:

$$\overline{\sigma} = \begin{cases} -m \times e^{j\theta} \times y & 0 \leq y \leq y_7 \\ -\frac{4}{3} \times e^{j\frac{\pi}{3}} \times y_7 + (\frac{4}{3} \times e^{j\frac{\pi}{3}} - m \times e^{j\theta}) \times y & y_7 \leq y \leq y_2 \\ -\frac{4}{3} \times e^{j\frac{\pi}{3}} \times y_7 + \frac{4}{3} \times (e^{j\frac{\pi}{3}} - 1) \times y_2 & \\ \quad +(\frac{4}{3} - m \times e^{j\theta}) \times y & y_2 \leq y \leq y_1 \\ -\frac{4}{3} \times e^{j\frac{\pi}{3}} \times y_7 + \frac{4}{3} \times (e^{j\frac{\pi}{3}} - 1) \times y_2 & \\ \quad +\frac{4}{3} \times y_1 - m \times e^{j\theta} \times y & y_1 \leq y \leq 1 \end{cases} \qquad \text{[A2.97]}$$

where $y_7 = \frac{t_7}{T_d}$, $y_2 = y_7 + \frac{t_2}{T_d}$, $y_1 = y_2 + \frac{t_1}{T_d}$ are coefficients imposed by projections of the reference vector onto the vectors used during the switching period (in this case, $\overrightarrow{V_1}$ and $\overrightarrow{V_2}$). Next, the switching period is filled in by applying the null vectors $\overrightarrow{V_0}$ and $\overrightarrow{V_7}$ (used for the same periods as in classic PWM).

The RMS value of vector $\vec{\sigma}$ over a fundamental period, which we will denote by $\psi_f$, may be calculated following the method explained in Appendix 1. The relationship between the RMS value of $\vec{\Sigma}$ over a fundamental period $RMS(\left\|\vec{\Sigma}\right\|)$ and $\psi_f$ is as follows:

$$RMS(\left\|\vec{\Sigma}\right\|) = \Sigma_0 \times \psi_f \qquad \text{[A2.98]}$$

In other words, $\psi_f$ is the normalized value in relation to $\Sigma_0$ of the RMS value of $\vec{\Sigma}$ over a fundamental period. In the following, $\psi_f$ will be used as a tool for comparing strategies.

Using the same PWM strategy, by increasing the switching frequency, we reduce the value of $T_d$, and consequently the value of $\Sigma_0$. In qualitative terms, this relates to the fact that current harmonics are reduced as the switching frequency increases. In order to compare different PWM strategies, the same switching frequency must be used in all cases; this comes down to comparing strategies using the value of $\psi_f$, which is independent of the switching frequency. Clearly, from this perspective, the best strategies will present the lowest value for $\psi_f$ (reduced ripple in load currents).

The value of $\psi_f$ for the SVPWM strategy may be obtained using the following formula [HAV 98]:

$$\psi_f(m) = \sqrt{\frac{3}{\pi}\left[\frac{\pi}{36}m^2 - \frac{2\sqrt{3}}{27}m^3 + \left(\frac{\pi}{32} - \frac{3\sqrt{3}}{128}\right)m^4\right]} \qquad \text{[A2.99]}$$

This analytical expression is particularly interesting as the SVPWM strategy is generally considered in published literature as a benchmark for the evaluation of other techniques.

Figure A2.5 shows its appearance as a function of $m$. This curve is applicable for all values $\varphi$ of the phase deviation between currents and voltages in a load.

## A2.5. Spectral analysis of the DC bus current

The method presented in section A2.2.3 allows us not only to calculate the spectrum of inverter output voltages but also the spectrum of the inverter input current. Bierhoff *et al.* [BIE 08] applied the method proposed by Black [BLA 53] for different PWM strategies, based on the double Fourier series decomposition. We will not go into detail concerning this work here, but useful information on this subject may also be found in [NGU 11a].

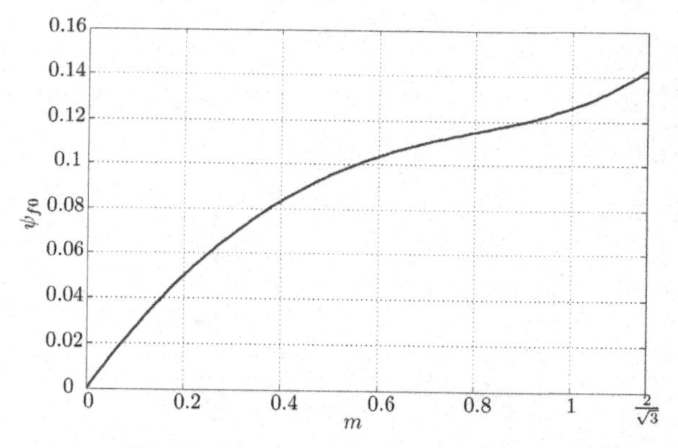


**Figure A2.5.** *RMS value of the normalized harmonic flow for the SVPWM strategy as a function of m in the linear zone*

# Bibliography

[APP 02] APPEL W., *Mathématiques pour la physique et les physiciens*, H & K, Paris, 2002.

[BAS 09] BASDEVANT J.-L., *Les mathématiques de la physique quantique*, Vuibert, Paris, 2009.

[BEC 00] BECH M.M., Analysis of random pulse-width modulation techniques for power electronic converters, PhD Thesis, Aalborg University, 2000.

[BEN 33] BENNETT W.R., "New results in the calculation of modulation productions", *Bell System Technical Journal*, vol. 12, no. 4, pp. 238–243, 1933.

[BIE 08] BIERHOFF M., FUCHS F.W., "DC link harmonics of three phase voltage source converters influenced by the pulsewidth modulation strategy – an analysis", *IEEE Transactions on Industrial Electronics*, vol. 55, no. 5, pp. 2085–2092, May 2008.

[BLA 53] BLACK H.S., *Modulation Theories*, Van Nostrand, New York, 1953.

[BOW 08] BOWICK C., BLYLER J., AJLUNI C., *RF Circuit Design*, Newnes/Elsevier, Boston, Amsterdam, 2008.

[BRÉ 05] BRÉHAUT S., Modélisation et optimisation des performances CEM d'un convertisseur AC/DC d'une puissance de 600 W, PhD Thesis, François Rabelais University, Tours, 2005.

[BÜH 91] BÜHLER H., *Convertisseurs statiques*, Presses Polytechniques and Universitaires Romandes, Lausanne, 1991.

[CHA 05] CHAROY A., *Compatibilité électromagnétique*, Dunod, Paris, 2005.

[CHE 09] CHEN L., PENG F.Z., "Closed-loop gate drive for high power IGBTs", *24th Annual Applied Power Electronics Conference and Exposition, APEC 2009, IEEE*, pp. 1331–1337, 2009.

[CHE 99] CHERON Y., *La commutation douce*, Tec & Doc, Paris, 1999.

[COH 00] COHEN DE LARA M., D'ANDRÉA-NOVEL B., Cours d'automatique – commande linéaire des systèmes dynamiques, Presses des Mines, Paris, 2000.

[COC 02] COCQUERELLE J.-L., PASQUIER C., *Rayonnement électromagnétique des convertisseurs à découpage*, EDP Sciences, Les Ulis, 2002.

[COR 13] CORNELL D., Aluminum electrolytic capacitor application guide, available at: www.cde.com/fliptest/ alum/alum.html, 2013.

[COS 05] COSTA F., MAGNON D., "Graphical analysis of the spectra of EMI sources in power electronics", *IEEE Transactions on Power Electronics*, vol. 20, no. 6, pp. 1491–1498, November 2005.

[COS 13] COSTA F., GAUTIER C., LABOURÉ E. *et al.*, *La compatibilité électromagnétique en électronique de puissance, Principes et cas d'études*, Hermès-Lavoisier, Paris, 2013.

[DEG 01] DEGRANGE B., *Introduction à la physique quantique*, Presses des Mines, Paris, 2001.

[ESC 87] ESCANÉ J.P., PACAUD A., *Synthèse des circuits passifs et actifs. Filtres*, Les cours de l'Ecole Supérieure d'Electricité, Eyrolles, 1987.

[FER 02] FERRIEUX J.-P., FOREST F., *Alimentations à découpages et convertisseurs à résonance*, 3rd ed., Dunod, Paris, 2002.

[FEY 99] FEYNMAN R., LEIGHTON R.B., SANDS M., *Cours de physique de Feynman, Electromagnétisme*, vol. 1 and 2, Dunod, Paris, 1999.

[FOC 98] FOCH H., FOREST F., MEYNARD T., "Onduleurs de tension – structures, principes, applications", *Techniques de l'Ingénieur*, Traité Génie Electrique, Article D3176, 1998.

[FOC 11] FOCH H., CHÉRON Y., "Convertisseur de type forward – dimensionnement", *Techniques de l'ingénieur*, Traité Génie Electrique, Article D3167, 2011.

[FRI 94] FRICKEY D.A., "Conversions between S, Z, Y, h, ABCD, and T parameters which are valid for complex source and load impedances", *IEEE Transactions on Microwave Theory and Techniques*, vol. 42, no. 2, pp. 205–211, 1994.

[GHA 03] GHAUSI M., LAKER K., *Modern Filter Design: Active RC and Switched Capacitor*, Noble Publishing, Atlanta, 2003.

[GIB 07] GIBSON W.C., *The Method of Moments in Electromagnetics*, Chapman & Hall/CRC, Boca Raton, 2007.

[HAV 98] HAVA A.M., KERKMAN R., LIPO T., "A high performance generalized discontinuous PWM algorithm", *IEEE Transactions on Industry Applications*, vol. 34, no. 5, pp. 1059–1071, 1998.

[HAV 99] HAVA A.M., LIPO T.A., KERKMAN R.J., "Simple analytical and graphical methods for carrier-based PWM-VSI drives", *IEEE Transactions on Power Electronics*, vol. 14, no. 1, pp. 49–61, 1999.

[HOB 05] HOBRAICHE J., Comparaison des stratégies de modulation à largeur d'impulsions triphasées – Application à l'alterno-démarreur, PhD Thesis, UTC, Compiègne, 2005.

[HOL 83] HOLTZ J., STADTFELD S., "A predictive controller for a stator current vector of AC-machines fed from a switched voltage source", *International Power Electronics Conference IPEC*, vol. 2, pp. 1665–1675, Tokyo, 1983.

[IEE 12] IEEE Standards Association, IEEE Electromagnetic Compatibility Standards Collection: VuSpec™, CD-ROM, 2012.

[KEM 12] KEMET, Electrolytic capacitors, Documentation technique, FF3304 06/09, available at: www.kemet.com, 2012.

[KOL 91] KOLAR J.W., ERLT H., ZACH F.C., "Influence of the modulation method on the conduction and switching losses of a PWM converter system", *IEEE Transactions on Industry Applications*, vol. 27, no. 6, pp. 399–403, 1991.

[KOL 93] KOLAR J.W., Vorrichtung und Verfahren zur Umformung von Drehstrom in Gleichstrom, IXYS Semiconductors GmdH, Patent no. EP0660498 A2, 23 December 1993.

[LAN 09] LANFRANCHI V., PATIN N., DÉPERNET D., "MLI précalculées et optimisées," in MONMASSON E. (ed.), *Commande rapprochée de convertisseurs*, Hermès-Lavoisier, Paris, pp. 113–137, 2009.

[LEF 02] LEFEBVRE S., MULTON B., "Commande des semi-conducteurs de puissance: principes", *Techniques de l'Ingénieur*, D-3231, 2002.

[LES 97] LESBROUSSARD C., Etude d'une stratégie de modulation de largeur d'impulsions pour un onduleur de tension triphasé à deux ou trois niveaux: la Modulation Delta Sigma Vectorielle, PhD Thesis, UTC, 1997.

[LIN 10] LINDER A., KANCHAN R., KENNEL R. *et al.*, *Model-Based Predictive Control of Electrical Drives*, Cuvillier Verlag, Gttingen, 2010.

[LUM 00] LUMBROSO H., *Ondes électromagnétiques dans le vide et les conducteurs, 70 problèmes résolus*, 2nd ed., Dunod, Paris, 2000.

[MAT 09] MATHIEU H., FANET H., *Physique des semiconducteurs et des composants*, 6th ed., Dunod, Paris, 2009.

[MEY 93] MEYNARD T., FOCH H., "Imbricated cells multilevel voltage-source inverters for high-voltage applications", *EPE Journal*, vol. 3, no. 2, pp. 99–106, June 1993.

[MIC 05] MICROCHIP, Sinusoidal control of PMSM motors with dsPIC30F3010, AN1017, available at: ww1.microchip.com/downloads/en/AppNotes/01017A.pdf, 2005.

[MID 77] MIDDLEBROOK R.D., CK S., "A general unified approach to modeling switching converter power stages", *International Journal of Electronics*, vol. 42, no. 6, pp. 521–550, 1977.

[MOH 95] MOHAN N., UNDELAND T.M., ROBBINS W.P., *Power Electronics – Converters, Applications and Design*, 2nd ed., Wiley, New York, 1995.

[MON 97] MONMASSON E., FAUCHER J., "Projet pédagogique autour de la MLI vectorielle", *Revue 3EI*, no. 8, pp. 22–36, 1997.

[MON 09] MONMASSON E. (ed.), *Commande rapprochée de convertisseurs statiques 1, Modulation de largeur d'impulsion*, Hermès-Lavoisier, Paris, 2009.

[MON 11] MONMASSON E. (ed.), *Power Electronic Converters: PWM Strategies and Current Control Techniques*, ISTE Ltd, London and John Wiley & Sons, New York, 2011.

[MOR 07] MOREL F., Commandes directes appliquées à une machine synchrone à aimants permanents alimentée par un onduleur triphasé à deux niveaux ou par un convertisseur matriciel triphasé, PhD Thesis, INSA de Lyon, 2007.

[MOY 98] MOYNIHAN J.F., EGAN M.G., MURPHY J.M.D., "Theoretical spectra of space vector modulated waveforms", *IEE Proceedings of Electrical Power Applications*, vol. 145, pp. 17–24, 1998.

[MUK 10] MUKHTAR A., *High Performance AC Drives*, Springer, Berlin, 2010.

[NAR 06] NARAYANAN G., KRISHNAMURTHY H.K., ZHAO D. *et al.*, "Advanced bus-clamping PWM techniques based on space vector approach", *IEEE Transactions on Power Electronics*, vol. 21, no. 4, pp. 974–984, 2006.

[NAR 08] NARAYANAN G., RANGANATHAN V.T., ZHAO D. *et al.*, "Space vector based hybrid PWM techniques for reduced current ripple", *IEEE Transactions on Industrial Electronics*, vol. 55, no. 4, pp. 1614–1627, 2008.

[NGU 11a] NGUYEN T.D., Etude de stratégies de modulation pour onduleur triphasé dédiées à la réduction des perturbations du bus continu en environnement embarqué, PhD Thesis, UTC, Compiègne, 2011.

[NGU 11b] NGUYEN T.D., HOBRAICHE J., PATIN N. *et al.*, "A direct digital technique implementation of general discontinuous pulse width modulation strategy", *IEEE Transactions on Industrial Electronics*, vol. 58, no. 9, pp. 4445–4454, September 2011.

[OSW 11] OSWALD N., STARK B., HOLLIDAY D. *et al.*, "Analysis of shaped pulse transitions in power electronic switching waveforms for reduced EMI generation", *IEEE Transactions on Industry Applications*, vol. 47, no. 5, pp. 2154–2165, October–November 2011.

[PAT 15a] PATIN N., *Power Electronics Applied to Industrial Systems and Transports - Volume 1*, ISTE Press, London and Elsevier, Oxford, 2015.

[PAT 15b] PATIN N., *Power Electronics Applied to Industrial Systems and Transports - Volume 2*, ISTE Press, London and Elsevier, Oxford, 2015.

[PAT 15c] PATIN N., *Power Electronics Applied to Industrial Systems and Transports - Volume 3*, ISTE Press, London and Elsevier, Oxford, 2015.

[REB 98] REBY F., BAUSIERE R., SOHIER B. *et al.*, "Reduction of radiated and conducted emissions in power electronic circuits by the continuous derivative control method (CDCM)", *Proceedings of the 7th International Conference on Power Electronics and Variable Speed Drives*, pp. 158–162, 1998.

[REV 03] REVOL B., Modélisation et optimisation des performances CEM d'une association variateur de vitesse – machine asynchrone, PhD Thesis, Joseph Fourier University, 2003.

[ROM 86] ROMBAULT C., SÉGUIER G., BAUSIÈRE R., *L'électronique de puissance – Volume 2, La conversion AC-AC*, Tec & Doc, Paris, 1986.

[ROU 04a] ROUDET J., CLAVEL E., GUICHON J.M. *et al.*, "Modélisation PEEC des connexions dans les convertisseurs de puissance", *Techniques de l'Ingénieur*, D-3071, 2004.

[ROU 04b] ROUDET J., CLAVEL E., GUICHON J.M. *et al.*, "Application de la méthode PEEC au cablage d'un onduleur triphasé", *Techniques de l'Ingénieur*, D-3072, 2004.

[ROU 04c] ROUSSEL J.-M., *Problèmes d'électronique de puissance*, Dunod, Paris, 2004.

[SCH 99] SCHELLMANNS A., Circuits équivalents pour transformateurs multienroulements, Application à la CEM conduite d'un convertisseur, PhD Thesis INPG, July 1999.

[SCH 01] SCHNEIDER E., DELABALLE J., La CEM: la compatibilité électromagnétique, Cahier technique, no. 149, 2001.

[SEG 11] SÉGUIER G., DELARUE P., LABRIQUE F., *Electronique de puissance*, Dunod, Paris, 2011.

[SHU 11] SHUKLA A., GHOSH A., JOSHI A., "Natural balancing of flying capacitor voltages in multicell inverter under PD carrier-based PWM", *IEEE Transactions on Power Electronics*, vol. 56, no. 6, pp. 1682–1693, June 2011.

[VAS 12] VASCAS F., IANNELI L. (eds), *Dynamics and Control of Switched Electronic Systems*, Springer, Berlin, 2012.

[VEN 07] VENET P., Amélioration de la sureté de fonctionnement des dispositifs de stockage d'énergie, mémoire d'HDR, University Claude Bernard – Lyon 1, 2007

[VIS 07] VISSER J.H., Active converter based on the vienna rectifier topology interfacing a three-phase generator to a DC-Bus, Master's Thesis, University of Pretoria, South Africa, 2007.

[VIS 12] VISSER H.J., *Antenna Theory and Applications*, Wiley, New York, 2012.

[VOG 11] VOGELSBERGER M.A., WIESINGER T., ERTL H., "Life-cycle monitoring and voltage-managing unit for DC-link electrolytic capacitors in PWM converters", *IEEE Transactions on Power Electronics*, vol. 26, no. 2, pp. 493–503, February 2011.

[WEE 06] WEENS Y., Modélisation des câbles d'énergie soumis aux contraintes générées par les convertisseurs électroniques de puissance, PhD Thesis, USTL, Lille, 2006.

[WHE 04] WHEELER P.W., CLARE J.C., EMPRINGHAM L. *et al.*, "Matrix converters", *IEEE Industry Applications Magazine*, vol. 10, no. 1, pp. 59–65, January–February 2004.

[WIL 99] WILLIAMS T., *Compatibilité électromagnétique – De la conception à l'homologation*, Publitronic, Paris, 1999.

[XAP 05] ALEXANDER M., Power distribution system (PDS) design: using bypass/decoupling capacitors, Application Note, XAPP623, available at: www.xilinx.com, February 2005.

[YUA 00] YUAN X., BARBI I., "Fundamentals of a new diode clamping multilevel inverter", *IEEE Transactions on Power Electronics*, vol. 15, no. 4, pp. 711–718, July 2000.

[ZHA 10] ZHAO D., HARI V.S.S.P.K., NARAYANAN G. *et al.*, "Space-vector-based hybrid pulsewidth modulation techniques for reduced harmonic distortion and switching loss", *IEEE Transactions on Power Electronics*, vol. 25, no. 3, pp. 760–774, 2010.

# Index

**G, H, I**



**K, L, M**



**N, P**



**Q, R**



**S, T**

**V, W, Z**